[英国]保罗·布拉斯利 理查德·索费 著　朱邦芊 译

牛津通识读本·

农业
Agriculture
A Very Short Introduction

译林出版社

图书在版编目（CIP）数据

农业 /（英）保罗·布拉斯利（Paul Brassley），
（英）理查德·索费（Richard Soffe）著；朱邦芊译. —南京：
译林出版社，2022.7
（牛津通识读本）
书名原文：Agriculture: A Very Short Introduction
ISBN 978-7-5447-9141-0

I.①农… II.①保… ②理… ③朱… III.①农业史 – 世界
IV.①S-091

中国版本图书馆 CIP 数据核字（2022）第 063015 号

Agriculture: A Very Short Introduction by Paul Brassley and Richard Soffe
Copyright © Paul Brassley and Richard Soffe 2016
Agriculture: A Very Short Introduction was originally published in English in 2016. This licensed edition is published by arrangement with Oxford University Press. Yilin Press, Ltd is solely responsible for this bilingual edition from the original work and Oxford University Press shall have no liability for any errors, omissions or inaccuracies or ambiguities in such bilingual edition or for any losses caused by reliance thereon.
Chinese and English edition copyright © 2022 by Yilin Press, Ltd
All rights reserved.

著作权合同登记号　图字：10-2018-121 号

农业　[英国] 保罗·布拉斯利　理查德·索费 / 著　朱邦芊 / 译

责任编辑　许　丹
装帧设计　景秋萍
校　　对　王　敏
责任印制　董　虎

原文出版　Oxford University Press, 2016
出版发行　译林出版社
地　　址　南京市湖南路 1 号 A 楼
邮　　箱　yilin@yilin.com
网　　址　www.yilin.com
市场热线　025-86633278
排　　版　南京展望文化发展有限公司
印　　刷　江苏扬中印刷有限公司
开　　本　890 毫米 × 1260 毫米　1/32
印　　张　8
插　　页　4
版　　次　2022 年 7 月第 1 版
印　　次　2022 年 7 月第 1 次印刷
书　　号　ISBN 978-7-5447-9141-0
定　　价　39.00 元

版权所有·侵权必究

译林版图书若有印装错误可向出版社调换。质量热线：025-83658316

序 言

陈 阜

"民以食为天",农业是人类文明的起源和社会经济发展的基础支撑,让更多的人了解和关注农业是非常必要的。农业从最原始的"种地养畜"到传统的"农林牧副渔",再到现代的"农工贸、产加销",农业的内涵及产业框架变得越来越庞大而且复杂。尤其在现代生物技术、信息技术、制造技术、装备技术日新月异,以及农业商业化、资本化、国际化飞速发展背景下,农业所涉及的一、二、三产业领域越来越广泛,农业发展路径与发展模式更加丰富多样。近年来,我国非常重视科普工作,陆续出版发行了一大批与现代农业相关的各种科普书籍,但针对农业整体性介绍的科普著作极少。同时,强化通识教育也是我国高等教育改革发展的重点,几乎全部涉农高校都开设了面向所有专业的"农业概论""农学概论"等通识课程,也编写了一大批农业通识教材,但普遍存在定义含义多、框架理论杂、概念套概念,过于注重对农业发展历程及产业体系、学科体系、技术体系的系统化介绍,以及需要大量识记才能理解掌握等诸多问题,容易让学生感到内容庞杂和枯燥乏味。由此可见，编写一本通俗易懂、生动有趣的农业通识读本并不是一件容易的事情。

　　第一次翻看译林出版社出版的"牛津通识读本"《农业》初稿，就感觉这是一本浅显易懂、可以轻松阅读的好书。它从人们都熟悉且天天面对的鸡蛋、牛奶、黄油说起，没有太多深奥难懂的概念、理论和技术介绍，甚至你不需要从头到尾地读，随便翻开一页就能读下去，这对普通读者简单了解农业及其发展非常有帮助。该书作者保罗·布拉斯利（Paul Brassley）和理查德·索费（Richard Soffe）是欧洲知名的农业专家，在农业和农村发展方面发表过许多有独到见解的论文、编著和书评，这本书用仅仅几万字描绘农业涉及的主要内容和基础知识，也充分体现了编著者宽阔的知识背景与突出的综合能力。这本小书有三个显著特点：一是非常简单但很准确地抓住了几个核心要素来介绍农业，这些内容包括土壤与农作物生产、饲料与畜禽生产、农产品贸易与国家粮食安全、农业投入与可持续发展、农业现代化进程与未来发展趋势；二是白描式地介绍了世界各地多种多样的农业类型及其差异特征，全球同时存在的现代农业、传统农业甚至原始农业的做法及效果，没有掺杂个人观点和过多的道理说教，正像作者强调的，"无论是科学耕作方法还是传统耕作方法，本书既不批评，也不打算为二者之一辩护。本书承认世界的多样性，在这个世界上发达国家和发展中国家的农业随处可见"；三是虽然作者没有明确提出当今农业发展必须重视和解决什么问题，但从开始描述的美国20世纪30年代"黑风暴"带来的土壤破坏，到灌溉、机械、农药和化肥大量使用可能带来的资源生态危机，再到未来农业的政策取向、气候变化影响、转基因生物及养活全球90亿人口的难题，都在提醒人类社会应该关注和思考这些问题和挑战。

当然，农业生产的地域性很强，不同国家和地区的资源禀赋与生态环境、社会经济状况与科技发展水平等差异很大，对农业发展道路和政策制定的选择也不可能相同。因此，每个读者对农业及其发展的感受、认识和观点肯定也不同，但这不影响对这部通识读本知识性和思想性的接纳。同时，也期望国内从事农业科普和通识教育的同行们能够从本书编写思路和方式上得到一些启发，编写出更多更好的我们自己的农业通识读本和教材。

目 录

致　谢　1

引　言　1

第一章　土壤与农作物　1

第二章　农场动物　27

第三章　农产品与贸易　44

第四章　农业投入　60

第五章　现代和传统农业　76

第六章　农业的未来　89

索　引　104

英文原文　111

致　谢

　　我们在职业生活中思考并与众多熟人、学生、同事和朋友讨论农业，他们都以这样或那样的方式影响了我们的观点。对于他们对本书手稿提出的有益意见，我们特别感谢安吉拉·布拉斯利、卡罗琳·布拉斯利、查尔斯·布拉斯利、查理·克拉特巴克、艾伦·库珀、史蒂夫·贾维斯、安妮塔·耶灵斯、罗布·帕金森、艾莉森·塞缪尔、牛津大学出版社的编辑安德烈娅·基根和珍妮·纽吉，以及我们的匿名读者们。我们非常感谢卡丽·希克曼在图表和照片方面所做的工作，还要感谢丹·哈丁细致、敏锐和专业的文字编辑。

引 言

想象你的面前有一个鸡蛋。农业研究的就是如何让母鸡下蛋。还有就是配鸡蛋吃的其他食物的原料生产，比如面包和黄油。

母鸡是从哪儿来的？当然是来自另一个鸡蛋，但在此例中，鸡蛋是由公鸡授过精的，鸡胚在蛋壳里发育。如今，工业化国家的大多数商业规模化鸡蛋生产商都从专业育种者那里购买产蛋鸡，这些育种者雇用遗传学家来培育更高产的母鸡，现在每只母鸡每年可以产300个或更多的鸡蛋。这大约是1950年产量的两倍。

得到母鸡后，农民就得给母鸡找个地方住。母鸡都是印度原鸡的驯化后代，它们本身还保留着野生祖先的一些行为，会找隐蔽的地方下蛋，晚上在树上栖息。因此，饲养母鸡数量不多的人，往往会在鸡舍里设置产蛋箱，里面放有筑巢材料和栖架，让母鸡在上面栖息。另一个极端是使用传统的层架式鸡笼，小笼子里会有两三只母鸡站在铁丝网地板上。自2012年起，这种

1 鸡笼在欧盟（EU）已被禁止了。取代它的"富集型"鸡笼体积要大得多，可以容纳更多的母鸡，通常在40～80只之间，并配有巢箱和沙浴区。英国当下生产的蛋里约有一半来自笼中的母鸡，其余大部分由散养的禽群生产。在美国，90%以上的鸡蛋是在层架式鸡笼中生产的，不过有一两个州——例如加利福尼亚州——已经禁止使用层架式鸡笼了。

农民还要保证母鸡有食物和水。母鸡是天生的清道夫。它们会吃种子、果实、青草和其他绿叶、蠕虫和昆虫，还会啄食小颗的砂粒，这些砂粒会进入它们的砂囊，以物理的方式帮助它们分解食物。农民必须给关在笼子里的母鸡提供转化为鸡蛋的全部营养物质：碳水化合物、蛋白质、脂肪、矿物质（尤其是蛋壳所需的钙）和维生素，在许多散养系统中也是如此。所有这些都会在饲料厂里组合成颗粒或粉料，大型家禽企业的饲料厂可能设在农场内，但更多的是在工厂区，这些厂区往往靠近海港，方便获得进口原料。碳水化合物大多来自小麦、大麦或玉米等谷物，这些谷物都是田间生长的，其中某些田地可能在数千英里之外。同样，大豆也是欧洲和北美的母鸡获取蛋白质与脂肪的主要来源之一，其中有很多产自拉丁美洲。因此，鸡蛋与许多动物产品一样，与耕作或农作物生产联系在一起。

耕农生产谷物、蛋白质和油料作物、马铃薯或（热带国家的）木薯及薯蓣等根茎类作物，以及组成动物饲料和人类大部分食物的所有其他植物与植物性原料。例如，为了种植谷物，他们借助强力机械处理土壤，把种子播种到其中，并施以粪肥和化肥，在工业化国家和许多发展中国家，种子如今是遗传物质的一种
2 非常成熟的科学来源。过去七十年，化学工业生产了种类繁多

的除草剂、杀真菌剂和杀虫剂，可以保护生长中的植物不受真菌和昆虫的侵扰，免于杂草的竞争。最后，大型机械收割作物时，可将用于供人类和鸡食用的谷物与植物的其他部分分离开来。

但是，尽管发达国家和某些发展中国家的情况是这样的，与此同时，仍有许多国家的农民弯着腰，用他们祖先几百年来一直使用的那种锄头来准备苗床。这些农民可能无法饲养母鸡，因为他们所在的地区流行一种叫作鸡瘟的疾病，而他们付不起为母鸡接种疫苗所需的那点费用。

芝加哥以南五十英里处使用大型机械生产玉米和大豆的综合农业公司，莫桑比克木薯地旁手拿锄头的妇女，稻田里的中国集体农场工人，以及在慕尼黑有正职、同时兼营奶牛场的德国家庭，他们从事的都是农业。这部通识读本打算找出他们活动的共同特点和决定他们具体工作的普适原则，解释他们之间为什么存在差异，并探讨他们的活动所引起的一些争议。无论是科学耕作方法还是传统耕作方法，本书既不批评，也不打算为二者之一辩护。本书承认世界的多样性，在这个世界上，发达农业和发展中农业随处可见。

第一章

土壤与农作物

土 壤

1934年5月,一场暴风从北美大平原掀起了一团黑色的尘土,将1 200万吨黑土倾倒在芝加哥市(图1)。两天后,暴风到达东部沿海地区,灰尘飘进了华盛顿白宫的窗户。这场灾难有益的一面是提醒人们,人类的生命要依赖区区几英寸深的脆弱表土,对农民来说,它和阳光一样重要。作为固定作物和为其提供营养的资源,它是一种包含成千上万生物体的活介质。在世界各地,土壤的类型和厚度对作物的种植都至关重要。本章的第一部分将介绍土壤是什么,它们有何不同,又是如何分类的,以及农民如何管理它们。

土壤是由空气、水、矿物质和有机质组成的复杂混合物,在很多时候,要历经数千年的演变。通常情况下,在状态良好的表土体积中,空气占25%,水也占25%,有机质则占其体积的5%左右。因此,体积不等的实际矿物质部分(从大石头到细小

图1　1930年代的美国沙尘暴：一幅风沙土的生动画面

的黏土颗粒）只占表土体积的45%，尽管其在土壤重量中的比例要大得多。

任何特定地点的土壤的类型都取决于母质①、气候、所在地的地形、生活在其中和其上的各种有机体，以及时间。有些母质是固态岩石，如花岗岩、砂岩、白垩和石灰岩，或者板岩，而其他的母质则是表层沉积物，如河流冲积物，即被河流带走后沉积在洪泛区的物质。冲积土或许是世上最肥沃的土壤，而形成于沙丘上的土壤在肥沃程度上往往处于相反的一端。黄土和沙丘一样，是由风输送的物质形成的，但在这种情况下，涉及的物质要细得多。它覆盖了北美和中国北方的广大地区，后者来自戈壁

---

① 形成土层的基本地质材料，一般是基岩、表层沉积或冰川沉积。（本书脚注均为译者注）

沙漠。此外,还有在潮湿条件下形成的泥炭土、在冰川融化后的沉积物上形成的冰碛土、来自最初沉积于海底的物质的海洋黏土(如荷兰部分地区),以及火山灰土。

这些和其他母质最终能发展出什么,则取决于当地的气候,特别是当时的温度和湿度、生活于斯的有机体,以及该地的地形,特别是坡度。坡度越大,反复冻融或暴雨就越容易使物质下移,所以缓坡地的土壤往往比陡坡地的更厚。气温的升高加速了矿物质的风化和有机质的分解,土壤湿度越大,就有越多的钙等可溶性元素从表层向下移动,所以湿润的热带或半热带土壤往往具有相当高的酸度,而温度相近地区的干燥土壤,其酸度则不那么高。土壤上生长的植物从土壤中获得水分和养分;在自然条件下,它们死亡后成为腐烂的表层枯落物的一部分,分解后的养分又可以被重新吸收。

在热带森林中,沉积在地表的枯落物数量可达北方松林的十倍,但更温暖条件下的分解速度过快,以至于热带土壤中的有机质含量可能仍然很低。部分植物残体的分解和混合进入土壤是由小动物完成的,如蚯蚓、马陆、弹尾虫、螨虫、线虫(鳗蛔虫)、蚂蚁和白蚁等。其他分解者有真菌、细菌、放线菌(像细菌一样是单细胞,但像真菌一样产生菌丝线)、藻类和病毒等。与之相对应的是脊椎动物,如兔子、鼹鼠、草原犬鼠、盲鼹鼠等,它们都是通过挖掘活动来促进土壤层混合的。当然,农民的耕种行为也会混合土壤。

这里有一个循环的过程,一般被称为碳循环:大气中的二氧化碳被固定在植物中(见后文中"植物如何生长"一节),这些植物要么死亡并被分解,要么被人类和其他动物吃掉,从而部分固定在其体内,残余物作为粪便和尿液出现,又被分解并成为土壤有机质的一部分,其中一些将被氧化,从而回到大气的二氧化碳库中。这一切发生的速度显然受到了农业发展的影响,因为放牧动物会去除生长中的植物,将其转化为粪尿,也会转化为气态二氧化碳和甲烷。它或许在更大程度上还受到了煤炭和石油形式的碳化石开采的影响,这增加了大气中的二氧化碳含量。

所有这些因素都有助于形成不同类型的土壤,我们将在后文中看到这些,但对农民个人来说,最重要的可能是土壤的质地和结构。质地是由土壤中不同大小的矿物质颗粒的比例决定的,如表1所示,该表显示了英国采用的分类系统。

不同的土壤或多或少都会含有这些不同的颗粒。例如,根据美国农业部的体系,由40%的沙子、40%的淤泥和20%的黏土组成的土壤,将被归类为"壤土",而这三种成分数量相当的土

表1 土壤矿物质的颗粒大小和性质

	直径(毫米)	特 征
石子或石头	大于2	—
粗沙	0.2～2	沙滩沙,颗粒可见
细沙	0.06～0.2	沙漏沙
淤泥	0.002～0.06	粉状物,只有在手持放大镜下可见
黏土	小于0.002	腻子状,只有在电子显微镜下可见

数据来源:R. J. 帕金森,《土壤管理与作物营养》,见R. J. 索费编,《农业手册》第20版,第4页。© 布莱克威尔科学有限公司2003,布莱克威尔出版公司。

壤则为"黏壤土"。要精确测定粒度分布,需要经过化学处理、筛分和沉积等复杂的过程,但有经验的土壤科学家可以通过一种被称为"吐唾沫然后揉搓"的过程,进行足够有效的评估。土壤在手指和拇指之间被润湿,通过对样品的感触便可了解其质地。粗沙让人感觉很有颗粒感,细沙则稍逊一筹,淤泥让样品变得柔滑,黏土则使其黏稠。

土壤的质地会影响到它的易耕性和肥沃度。含沙量高的土壤容易排水,但也会受到干旱的影响,不是特别肥沃。耕种这种土壤的人经常说,他们每天都需要下雨,星期天则要淋粪肥,虽然他们或许不会说得如此确切。而黏土则要肥沃得多,但耕作起来也困难得多。

有此区别的原因是什么?沙子和淤泥级别的土壤颗粒每克表面积相对较小,主要由石英(二氧化硅)颗粒组成,后者的化学性质不活泼。因此,它们不太可能通过毛细引力和化合物中的养分来保持水分。而一定重量的黏土颗粒具有巨大的表面积,往往比沙子的表面积大一千倍,并且化学性质活泼,因而能保持水分和养分。所以,它们可能非常肥沃;但另一方面,在黏土里拉动犁或其他任何中耕机都比沙土中需要更大的力量,而且黏土排水缓慢,干旱期难以复湿。因此,黏土常被形容为又"冷"又"重",而沙土则被认为是"轻盈的"。当然,大部分土壤既没有过多的沙子,也没有过多的黏土,所以既肥沃又容易耕作,壤土或黏土等从农耕的角度来看最好的土壤无疑也是如此。

农民可以通过人工排水来影响土壤的含水量,从而改善土壤的可耕性,延长生长季节,提高肥料的使用效率,刺激作物扎根更深。在英国,对农作物产量的影响在10%～25%之间。采用的技术多种多样,既有通向溪流及河水的明沟,也有瓦渠、管渠和鼠道渠。在鼠道渠这个例子中,所谓的鼠道是一个直径约70毫米的尖圆筒,连接在垂直犁片的底部,由强力拖拉机拉过土壤。在黏土中,这样产生的渠道可以维持长达五年之久。由专业机器铺设的瓦渠或管渠排水管的使用寿命要长得多,但安装费用也相应较高。

与此相反的过程当然是灌溉,这在某些国家里已经进行了几百年甚至几千年:尼罗河三角洲的季节性洪灾就是一个明显的例子。在温带国家,往往只有马铃薯、甜菜和各种蔬菜等价值较高的作物才能回报灌溉的费用,但在热带地区,灌溉是水稻作物种植的必要组成部分。据估计,全球可耕种面积的20%目前要进行灌溉,其农作物产量约占总产量的40%。大部分这类耕地种植的是亚洲水稻作物,但美国也有很多。

对于农民来说,另一个重要的考虑因素是土壤的结构,因为这也影响到耕作的容易程度。土壤的形成不是沙子、淤泥和黏土简单地混合在一起就行了,而是需要随着时间的推移,通过土壤生物的活动,将前面讨论的矿物质颗粒与有机质相混合,结果便形成了稳定的聚集体,其中含有植物根系和其他土壤生物所需的水和空气(图2)。

孔隙空间中可以充满水或空气,或是两者的混合物,在结构良好的土壤中,表层土壤颗粒均匀,越深处则聚集体越多。在这样的土壤中,栽培工具比较容易产生结构细密的表层土壤,农民称之为"耕性良好",这为植物的发芽和生长提供了理想的条件。相反,在结构不佳的土壤中,如在过于潮湿时被耕作过的黏壤土,黏土会变成"一片泥糊",或聚集成大块,往往在25～30

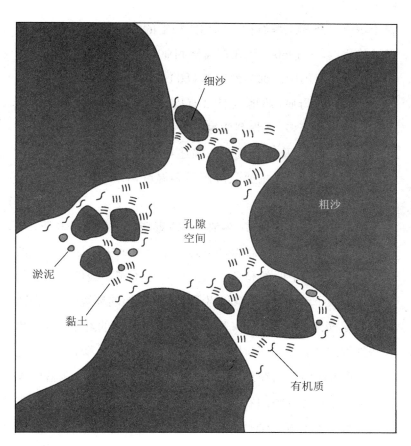

图2　土壤结构详解

厘米深的地方形成不透水的磐层,阻碍了排水。

农民无法改变土壤的质地(即沙子、淤泥和黏土的百分比),但他们的农场管理决策可以影响土壤的结构。连续耕作——或者在错误的情况下,即便是常规的耕作——会降低土壤中有机质的含量,更容易产生这种结构问题,而使用永久牧场或者至少是长期的草地,以及定期添加粪肥和石灰,都有助于改善土壤有机质和土壤结构。结构在极端的情况下会恶化到易受风蚀的程度，如1960年代在东盎格利亚①发生的情况，而最著名的例子就是美国中西部的沙尘暴（图1）。

所有这些母质、质地、结构、有机质和地形的变化意味着世上的土壤种类繁多，土壤科学家已经为其制定了各种分类系统，其中许多系统涉及极其复杂的描述性术语。最广泛使用的系统与气候变量有关，主要是降雨量和温度，以及由此产生的植被区。

因此，褐土（有时称之为雏形土）形成于温带落叶林下，是良好的混合耕作土壤，而黑钙土是在干草原或大草原上形成的黑土，成为美国中西部和乌克兰的可耕作土壤。酸性浸出型的灰化土形成于针叶林、荒原和沼泽地寒冷潮湿的环境中，而在热带地区温暖潮湿的环境中，则发现了高度风化的红土（也被称为砖红壤或铁铝土），有时呈红色或黄色。它们是酸性的，缺乏养分，虽然它们可以在地表快速回收利用养分，长出茂盛的森林，但只能成为贫瘠的农业土壤。灰黏土是积水土壤，在温带地区需要排水和精心管理，但在热带地区则是主要的水稻种植土壤。

这些只是以气候为基础的分类系统所识别的各种土壤中的一小部分，还有其他一些方法更强调母质和地形的重要性，从而形成了与地理和土地形态有关的全球分类。

因此，土壤从根本上支持着作物的生长，并为其根部提供水分。溶解在水中的是植物所需的各种养分，以帮助植物从空气

---

① 东英格兰一地区。

里的二氧化碳中提取其主要成分碳,我们将在下一节详细探讨这些养分。

作物的养分

植物的主要构成元素——碳、氢和氧——均来自它们通过叶子吸收的空气和根部吸收的水。除此之外,它们还大量需要三种元素(主要营养素),较少量需要另外三种元素,极少量需要通常被称为微量营养素的其他八种元素。

三种主要养分是氮(N)、磷(P)和钾(K)。大部分氮元素以硝酸根离子的形式从土壤水中获得,是植物蛋白质和叶绿素的组成成分。没有它,植物就会变黄,无法茁壮成长。归根结底,植物体内的氮来自空气,但它首先要进入土壤。少量的氮溶解在雨水中,但更多的氮来自生活在土壤中的藻类或细菌(如固氮菌)的活动,或是来自豆科植物(如豌豆、菜豆、白车轴草,甚至荆豆)根部的固氮结节。来自这些有机来源的氮被进一步的细菌活动转化为硝酸根离子,硝酸根离子极易溶于水,并以这种形式被植物根部吸收。不过,它们也可能随着水的流失而被溶出,或再次转化为大气中的氮,或者被土壤生物吸收,再次成为土壤有机储备的一部分。大多数粪肥和矿物质化肥的作用,就是增加这一循环中硝酸盐阶段氮的可利用量。

植物的能量和蛋白质代谢需要磷,没有磷,根就长不好。土壤中通常有大量的磷,但以相对难溶解的形式存在,所以在某些情况下,只有溶解状态的少量磷能被植物利用。钾参与植物的水分控制和运输机制,所以没有它,生长速度就会降低。钾在土壤里的黏土风化过程中被释放出来,由于水溶性高,可能会在排水中流失。在正常情况和大多数耕作制度下，只有这三种主要养分的流失量足够大，才需要用肥料来替代。

然而，在更集约化的农场管理系统中，还有一些其他的元素有时可能需要替换。硫是某些氨基酸和酶的成分，在正常情况下，土壤储量充足，但自从欧洲燃煤发电站的硫排放量下降后，一些地区出现了缺硫的迹象。这一类的另外两种元素是钙和镁，它们在控制土壤酸度方面起着一定的作用（见下一段）。最后是微量元素：硼、铜、锌、锰、钼、铁、氯和钴，其中许多元素在植物酶中起着重要作用，但通常需要的量很少，以至于大多数土壤都含有足够的量。例如，产生6吨谷物和3.5吨稻草的谷类作物将从1公顷土地上移走120公斤的氮，但导致的锌损失还不到1公斤。

土壤酸度也会影响养分的可利用性。它是由土壤的pH值来衡量的，pH值为7时，既不酸也不碱，随着pH值的降低，酸度会增加。耕地土壤的pH值通常在5～7之间，但对于像在石灰质和白垩上的钙质土壤来说，pH值可能在7～8之间。酸度最高的高地泥炭土可能低至3.5。在这个酸度水平上，所有主要营养素和某些微量营养素的可利用性都会降低。在pH值介于6～7之间的微酸性土壤中，养分的可利用性往往处于最高水平，微生物的活动也是如此，因此农民认识到需要通过添加适当的石灰质物料（如石灰石细粉）来维持这一水平。石灰也有助于形成良好的土壤结构，因此应将其视为土壤改良剂而不是肥料。

粪肥和化肥

从前面的讨论中可以看出，使用粪肥和化肥的目的是替代种植和放牧所带走的植物养分,所以它们的价值主要在于它们所含的氮、磷和钾。粪肥和化肥之间并没有硬性的界限。从本质上讲,粪肥的来源是有机的——典型的例子是农家肥,它是秸秆、粪便和尿液的混合物,而化肥的原料则是化工厂生产(如氮肥)或开采的(如磷肥和钾肥)。但也有很多介于这两个极端之间的产品的例子,比如血粉和骨粉,或者干鸡粪,它们源自有机质,但又经过了制造加工过程。

粪肥的养分含量相对较低:即使是最浓缩的家禽粪肥也只含有约2%的氮和磷以及1%的钾。农家肥的另一个好处是它含有部分降解的有机质,有助于改善土壤结构。近年来,高度机械化的农场将动物粪尿以浆液的形式收集起来,然后直接施用于土地而不添加秸秆,因此,这将不会产生同样的土壤结构效益。

有机肥(相对于粪肥)浓度更高,蹄角粉和血粉中含有12%～14%的氮。过去使用的此类原材料范围很广:海鸟粪、煤气石灰、鱼、鲸脂、"油渣"(蜡烛制造商的废料)、毛皮加工者的剪裁边角料、羽毛、羊毛、亚麻布、翻造呢绒、皮货商的鞣皮废料(尸体残骸)和菜籽粕粉都出现在1860年编制的清单上。同样,从煅烧石灰到泥灰岩和贝壳砂的许多原材料,过去都曾因其含有石灰而被使用。

无机肥的浓度最高,可利用养分最多能占施用量的一半。氮肥有好几种,如硝酸铵、尿素、无水氨等,但各种氮肥中的氮都是从大气中固定下来的(这个过程需要大量的烃类燃料,现在通常以气体的形式存在),然后再与其他元素结合。同样,大多数磷肥都来自磷灰岩,它可以直接作为化肥使用,但释放磷的过程非常缓慢。因此,通常用硫酸(制成过磷酸盐)或磷酸(制成磷含量更高的重过磷酸盐）来处理磷灰岩。

钾以氯化钾的形式被开采出来，要么直接使用，要么用硫酸处理后制成硫酸钾。所有这些单独的养分都可以独自使用——在这种情况下，它们通常被称为"单质肥"——但它们也可以结合起来制成"复合肥"，其中含有混合的养分，有时是为了满足特定作物的要求而配制的。因此，一种好的通用复合肥配比是20：10：10，含氮20%，磷和钾各占10%，而一种旨在替代因为收割青贮饲料而带走的养分的肥料不需要磷，因此比例可能是25：0：15（即25%的氮和15%的钾）。其中任何一种在特定的田地或作物上的施用量都取决于各种考虑因素，包括土壤中现有的氮、磷、钾状态到作物的反应、售价，以及肥料的成本。大多数作物的反应曲线是递减的，即随着施肥量的增加，作物产出的增产效应会降低。

因此，作物和草地的长期持续生产需要健康的、管理良好的土壤。一些农民认为，最好的办法是避免使用矿物质肥料，依靠有机方法循环利用养分。另一些人则采用精耕细作的方法，只在需要的地方投放矿物质肥料，以保持土壤肥力。农民必须根据他们现有的资源和他们希望种植的作物来做出正确的管理决策。

作　物

人类以植物为生，要么直接食用植物，要么间接食用动物的肉或产品（如牛奶），而这些都是动物进食植物后出产的。黑莓或护肤桐果等少数植物或许可以从野外采集，但大多数是栽培的，在这种情况下，它们被称为农作物。某种淀粉类能源作物几

图3 最重要的食用谷物：(a) 小麦，(b) 谷子，(c) 水稻，(d) 玉米

乎是世界各地人类饮食的基础，它通常是谷物，如小麦、谷子、水稻或玉米（图3）。在条件不适合种植这些作物的地方，则种植木薯、薯蓣或马铃薯等根茎类作物。本章的以下几节将探讨植物如何生长，其生长的阻碍因素，以及如何改善它们的生长，然后再介绍一些主要的作物和栽培制度。

植物如何生长

农民如何捕捉阳光中的能量，并使之可用，从而使诸位读者有力气翻开这本书的书页？他们利用了含有叶绿素的植物。叶绿素是让植物变绿的东西，它能吸收辐射，从而提供能量，推动叶子内部的一系列生化反应，从空气中吸收二氧化碳（CO_2），使之与根部吸收的水结合，制造简单的碳水化合物分子——

糖——并释放氧气（O_2）。这个过程叫作光合作用，以公式概括如下：

$$6CO_2 + 6H_2O \xrightarrow{\text{光}} C_6H_{12}O_6 + 6O_2$$

17　　它的重要性无论怎样强调都不为过。没有它，就没有我们所知的地球上的生命。从最矮小的草到最高大的树，所有的绿色植物都会进行光合作用，它们95%的干物质（即不含水的部分）都是由它产生的。葡萄糖等单糖可以结合成长链（称为聚合物），形成淀粉（大多数植物储存能量的形式）和纤维素（植物细胞壁的主要成分之一）。脂肪和油类也可以通过碳水化合物的进一步转化而产生，如果加入氮，有时再加入其他元素，还可以产生蛋白质。

　　因此，作物的生长速度显然取决于光合作用的速度，而后者的速度又取决于光照、温度、二氧化碳的浓度以及水和养分的供应情况。当温度和光照强度增加时，只要有水，光合作用也会加速。大多数温带作物使用所谓的C_3代谢途径进行碳固定，限制因素是大气中的二氧化碳浓度。这就是燃烧化石燃料导致这种气体水平的增加会产生肥效的原因，也是一些种植者提高温室中的二氧化碳含量的原因。

　　玉米、甘蔗、谷子和高粱等许多热带作物都使用另一种C_4代谢途径，在这种途径中，光合作用在光照强度大大提高之前不受二氧化碳浓度的限制。采取这种代谢方式的植物在更炎热干燥的条件下可以有很高的光合速率，因此人们目前正在着手研究确定起作用的基因，并利用它们来生产C_4品种的水稻。据称，这可以比目前的C_3品种多产50%的粮食。

随着植物的生长，由于细胞数量增加，干物质或总生物量也随之增加。植物通过细胞的特化（分化）和组织化，从种子萌芽到营养生长发育、繁殖发育，再到种子发育。根据其生命周期（一年生、两年生或多年生）以及植物有用的部分和作为食物收获的部分，作物可能在几周、几个月或几年后达到衰老（死亡）。

在发芽之前，大多数种子都能抵御寒冷和干旱的压力，往往可以长期存活。例如，斯瓦尔巴全球种子库①中的种子就保存在一口废弃矿井中温度零下的永久冻土里。大多数温带作物至少需要温度在 1～5 ℃才有可能发芽，最适宜的发芽温度在 20～25 ℃之间，但水稻的最适宜温度是 30～35 ℃。在植株生长阶段，植物的生长点（分生组织）产生叶片和茎秆材料，在生殖阶段产生花，花经过授粉受精后产生种子。对于粮食作物，农民只希望有足够的茎叶发育，最大限度地增加植物的光合（绿色）面积，因为他们将收获和出售的是种子，而他们根本不希望马铃薯产生种子，因为收获的是作为食物的块茎。

随着温度的升高，大多数作物生长得更快，原产于温暖气候下的玉米等作物，通常比原产于世界较冷地区的小麦或黑麦草等作物需要更高的温度，才能达到最佳生长状态。温度驱动着发育速度，并往往控制着从一个生长阶段到另一个阶段的变化时间。例如，一些小麦品种被设计为在秋季播种（冬小麦），需要一段时间的寒冷天气（春化作用）才能在来年夏天结籽。有些植物也会对日长反应灵敏。例如，一些水稻品种在白天开始变短之前不会开花，而大多数温带植物是"长日型"植物，需要

---

① 挪威政府在北冰洋斯瓦尔巴群岛上建造的非营利储藏库，用于保存全世界的农作物种子，是全球最大的种子库。

日长日益增长才能开始开花,这确保了它们在夏季开花。

这些都是单株植物的特点,但作物是植物的集合,个体之间的相互作用也很重要。农民力求对作物进行整体管理,在白昼最长、阳光最强的时候,最大限度地截取光照。因此,如果播种太晚,就可能无法产生足够的叶子来覆盖地面,播种太稀也会有同样的结果。肥沃的土壤和充足的水分也会促进叶子的生长。另一方面,如果这些因素中的任何一个过头了,都会导致产量下降,因为较高的叶子遮住了较低的叶子,单个植株没有足够的生长空间,或者由于快速生长或缺水,产生弱茎或垂茎,不能保持直立。

必须在单株植物的产量和作物整体的产量之间取得平衡。需要农民的技能和经验以及农业科学家的研究来确定最佳的种子、施肥量及播种时间,从而最大限度地提高作物的产量。

植物生长的阻碍因素

从植物和作物的这些特点可以看出,任何阻碍单株植物和作物整体以其最大的光合效率运作的因素都会降低收获和产出。从理论上讲,一般的作物可以捕捉到它可利用的大约5%的太阳能量;在实践中,这个数字接近1%,部分原因在于气候因素的影响,如寒冷或阴天,或是由于缺乏水分或养分,但主要是由于杂草、虫害和疾病的影响。

诚如谚语所说,杂草是长错了地方的植物。罂粟花(罂粟属)在花园里看起来很漂亮,但长在小麦作物中却是个问题。白茅(*Imperata cylindrica*)在马来语中被称为 lalang,在美国被称为 cogon,在澳大利亚则叫 blady grass,是东南亚传统的葺屋顶材料,也有栽培用于加固海滩的,在非洲有时也会用于放牧,但在美国东南部的几个州却是官方铲除计划的对象。杂草的影响还取决于它的生长时间:早期的杂草侵袭会使作物窒息,而后期在发育迅速的作物中生长的杂草对于产量可能就几乎没有影响了。因此,杂草最明显的影响来自它们争夺阳光从而减少作物的光合作用的能力,但它们也会与作物争夺水分,其种子可能会污染收获的作物,而且它们可能会使收获变得困难,例如旋花类植物爬上作物茎秆的时候,就会产生这样的问题。

因此,如果在错误的时间生长在错误的地方,几乎任何植物都可能是杂草,但有些被认为特别容易造成问题。在温带作物中,大穗看麦娘和野燕麦等禾本科杂草、一年生阔叶杂草和多年生杂草尤其难以对付。例如,藜(*Chenopodium album*)现在通常被归类为一年生阔叶杂草,但过去它曾被用作沙拉蔬菜。多年生的杂草包括匍匐冰草、欧洲蕨、酸模,以及温带地区的蓟;在热带地区,最麻烦的杂草之一是凤眼蓝(*Eichhornia crassipes*),它是作为观赏性植物从巴西引进的,但长成了野生植物并繁衍到堵塞灌溉渠道、引发洪水的程度。

和杂草一样,害虫/害兽则是走错了地方的动物。在塞伦盖蒂[①]平原上看起来很有气势的大象,践踏农作物时就是一种害兽;在林间草地上啃草的兔子可能会受到欢迎,但在树篱的另一侧啃食幼麦时就不是了。在大小尺度的另一端是微小的鳗蛔虫(线虫这种叫法非常合适),它们可以毁坏甜菜和马铃薯,还有包括蚜虫和蝗虫在内的许多昆虫。它们可以攻击作物从种子到

---

① 非洲坦桑尼亚西北部至肯尼亚西南部的地区。

根部的所有部分。有的通过吃植物的叶子来影响后者的光合能力,有的通过吃植物的根来影响植物的水分和养分吸收。蚜虫有刺吸式口器,以植物细胞中的汁液为食,所以令其生长受阻。在此过程中,麦长管蚜等有些蚜虫会将病毒性疾病(这里指大麦黄矮病毒BYDV)传给植物,进一步攻击后者。

当致病生物被引入新环境时,就会发生一些最具破坏性的害虫/害兽攻击。19世纪中叶之前,美国的科罗拉多马铃薯象甲一直在茄科的杂草上安静地生活着。后来引进了同科(茄科)植物马铃薯,象甲种群便增加并扩散开来。1873年,科罗拉多马铃薯象甲从科罗拉多州到达美国大西洋沿岸,1901年越过大西洋,并从那里传播到欧洲大陆,成为马铃薯的主要害虫。同样,美国的葡萄藤对根瘤蚜虫相对免疫,但当后者在19世纪传播到欧洲大陆的葡萄园时,却造成了巨大的破坏,因为那里使用的葡萄品种从未接触过根瘤蚜虫,没有进化出与美国品种相同的保护机制。

虫害的范围非常大,如果环境条件合适的话,几乎每一种作物上都会生活着几种害虫。例如,非洲的咖啡作物会受到丽蝽、盲蝽、咖啡果小蠹、粉蚧、介壳虫、蛀茎虫和潜叶虫的攻击。据估计,在整个世界范围内,虫害造成的作物损失高达25%。

农作物的病害多是由各种真菌引起的,但病毒和细菌也难辞其咎。有些真菌会造成叶绿素的损失,使叶子看起来发黄,因此减少了光合作用。另一些真菌则生长在植物输送水分的导管中,令植物枯萎。目前造成相当大问题的一个真菌的例子就是尖孢镰刀菌,它生活在土壤中,并渗透到香蕉的根部,最终蔓延开来,导致叶子变黄,整株植物开始枯萎。

尖孢镰刀菌是19世纪末首次在巴拿马发现的,因此也被称为"巴拿马病"。当时在香蕉商贸中占主导地位的品种"大麦克"(Gros Michel)对这种病很敏感,到1960年,该病已蔓延到大多数生产国。幸运的是,另一个品种"卡文迪许"被证明对当时存在的尖孢镰刀菌的菌种具有抗性,因此它继而成为主要的商业品种,但近年来马来西亚出现了一个被称为"热带菌种4号"(TR4)的新菌种,并传播到其他国家,而"卡文迪许"对它很敏感。由真菌致病疫霉引起的马铃薯枯萎病也许是植物病害中最著名的例子,它在19世纪中期袭击了爱尔兰的马铃薯作物,造成了巨大的社会经济影响。

该如何改善植物的生长状况?

如果植物因缺乏光、热、水和养分而无法生长,显而易见的解决办法就是提供这些东西,而且往往可以做到这一点,但要付出一定的代价。如果作物的价值足够高,这个代价或许是值得的。18世纪,一些富有的英格兰地主建造了加热的温室,在其中种植菠萝。近年来,警方在突击检查中发现了完全人工的种植室,毒贩在其中种植大麻。

在更常见的商业环境下,北欧的番茄种植者在温室中种植作物,人工加温、浇水、施肥,在某些情况下还增加了二氧化碳的含量。而在西班牙南部,同样的作物不需要人工加温,但需要从地下水井中抽取更多的水。在加利福尼亚州干旱地区上空飞行的旅客可以俯瞰到沙漠中的绿色圆圈,那是在环绕中央水源的大型灌溉设备的帮助下生长的作物(图4)。

如果运输成本足够低廉,那么在温暖的地方种植作物,然后

图4 中央枢纽灌溉促进了加利福尼亚州干旱地区的作物生长

把它出售到那些由于过于寒冷而无法生长的地方,可能是值得一试的:正因为如此,冬月里才能在欧洲超市买到肯尼亚的青豆。各地的农民总是试图以动物粪肥的形式将有机养分返还给土地,发达国家的大多数农民(有机农除外)以及发展中国家越来越多的农民现在则使用人工肥料(见本章前文)。

正如我们所看到的那样,即使在理想的生长条件下,作物的生长仍然可能会受到杂草和病虫害的影响,农民的技能之一就是掌握关于如何减少它们的影响的知识。当杂草与其周围生长的作物非常相似时,它们是最难控制的,无论通过耕种还是使用除草剂都很难根除。如果杂草看起来像作物,人们就会犹豫是否要用锄头处理掉它;如果它在生理和解剖上与作物相似,就很难找到一种既能杀死它又不损害作物的除草剂。例如,野生稻(*Oryza rufipogon*)是水稻中最严重的杂草,旁遮普地区的损失一度估计超过作物的一半。

农民已经制定了许多杂草防治的策略。重要的是以适当的播种率将未受污染的种子播入肥沃的苗床,使作物与杂草竞争获胜的概率最大化;有些杂草可以通过反复切割来控制;轮作可以使在阔叶作物中难以控制的阔叶杂草在随后种植的谷类作物中得到更有效的处理,反之亦然。

在采用化学除草剂之前,农场主及其工人们会用锄头清理作物,或者用手拔除杂草,对于杂草侵袭相当轻的高价值种子作物,现在有时仍会这样做。条播机可以成行播种,正如杰斯罗·塔尔[①]在其《马耕畜牧》(*Horse-Hoeing Husbandry*,1733年)一书中所论述的那样,发明这种机器的一个原因是,当时用马拉锄头在行间锄草比较容易,而易于实施行间耕作的宽行距作物通常被称为"抑草作物"。但所有这些方法都需要大量的劳动力。

20世纪初,在化学除草剂出现之前,一位农场主估计,一个农场有三分之一到一半的田间劳动都耗费在了消灭杂草上。因此就不难理解农场主为何如此热衷于使用除草剂(在时薪很高的经济体中尤其如此),也不难理解不能使用除草剂的有机农为何往往认为杂草比病虫害更难管理了。

如同杂草一样(病虫害也是如此),农民尝试用管理和化学方法相结合的方式来减少其影响。轮种作物使得土地在连续几年里生长不同的作物,有助于打破真菌和土壤中的昆虫的生命周期,还有通过早播或晚播来错过病虫害最活跃的时间,这些都

---

① 杰斯罗·塔尔(1674—1741),英格兰伯克郡的农业先驱,为英国农业革命做出了贡献。

是管理方法的例子。只有在这些方法无效时,农民才越来越多地使用昂贵的农药。

化学防治本身并非没有问题。它不仅可能使目标生物体对其产生抗药性,而且还可能对非目标物种产生不必要的副作用。广泛使用氯代烃杀虫剂对猛禽类繁殖成功率的影响就是一个众所周知的例子。解决这个问题的一个办法是利用捕食性动物打击害虫,如利用寄生蜂打击温室粉虱;另一个办法是尝试为植物培育出抵抗力。

抗病虫害只是植物育种者的潜在目标之一。从根本上说,大多数育种计划都试图提高作物产量或质量,或两者兼而有之,不过有时增产是以牺牲质量为代价的:例如,菲律宾的国际水稻研究所1966年发布的早期高产水稻品种之一IR8,由于谷粒呈粉质以及易碎而被消费者所厌弃。

育种者可以通过各种方式来提高产量。他们可以培育出抗旱或抗常见病虫害(如谷物叶片病)的品种,还可以尝试增加植物的可售部分,而牺牲其他部分。这已被水稻和小麦育种者广泛应用,他们培育出了秸秆较短但穗较大的品种。总的生物量可能不会有太大的变化,但可出售的生物量增加了。

其他的目标取决于单个作物的特质。老的甜菜品种生产的种子是一簇一簇的,发芽后会产生三到四棵植株,必须费力地用手锄从这些植株中挑出一棵强壮的,所以当育种者生产出"单胚芽"种子,每簇只有一粒种子时,就不需要这项工作了,成本也随之降低。糖的另一来源甘蔗的某些品种叶缘锋利,蔗田里的工人干活时可能会很疼,所以育种者试图生产出能尽量减少这一特性的品种。植物的现代育种方法及其影响将在第六章中讨论。

作物与种植

"我们大多数人的体内都有一个马铃薯形状的空间，每顿饭都必须填满它，不是用马铃薯就是用同样能填饱肚子的东西。"凯瑟琳·怀特霍恩在她的经典之作《床头柜里的烹饪》（*Cooking in a Bedsitter*，1963年）中写道。这些主食通常以淀粉为基础，来自一些最广泛种植的农作物，如水稻、小麦、玉米、马铃薯、木薯和薯蓣等。科幻作家约翰·克里斯托弗在其小说《草之死》（*The Death of Grass*，1956年）中提出，如果某种疾病开始杀死所有的禾本科植物，世界文明就会崩溃。这个科的植物不仅包括世上种植最广泛的作物——牧草，还包括温带谷类（小麦、大麦、燕麦、黑麦）、热带及半热带谷类（玉米、水稻、谷子、高粱）和甘蔗。如果类似的瘟疫影响到了包括豌豆、菜豆和兵豆在内的荚果（豆科），十字花科（卷心菜、油菜、芜菁甘蓝、芥菜等）也受到了打击的话，蔬菜和油菜籽等就都会灭绝，我们也会缺乏蛋白质。

有些农作物生产不止一种产品。小麦可以制作面包，大麦制成麦芽后可以制作啤酒，但这两种作物也可以用来喂养动物，特别是猪和鸡。大豆可以榨油，剩下的就是高蛋白动物饲料了。亚麻秆的纤维是制造亚麻布的来源，而亚麻籽则是工业油的来源，尽管随着时间的推移，不同品种的亚麻已经被培育成专门生产两种产品之一的作物了。

大多数农作物都有不同的品种，例如，小麦就有数百个不同的品种。有些品种经过培育，在炎热干燥的气候下生长得更好，有些品种能生产更好的面包面粉，还有些品种能抵抗各种真菌病害，等等。在热带和半热带国家种植的玉米品种构成了膳食淀粉最重要的来源之一，但这些品种不会是西北欧农民为生产动物饲料用的玉米青贮而种植的品种，也不会是美国农民为生产甜玉米而种植的品种。生产优质葡萄酒的葡萄品种并不适合作为新鲜的食用葡萄。

这些都是人类或动物的食用作物，但也有棉花、黄麻和大麻等大面积作物是为了获得纤维而种植的，还有小面积的琉璃苣、金盏花和月见草，用于生产工业用油和药用油。越来越多的作物可用于生产生物质，或是生物乙醇等燃料。

所有这些物种也可根据其生长方式进行分类。咖啡、可可、香蕉、橄榄和苹果等木本作物一旦种植后就能生长很多年。其他的多年生作物包括葡萄藤、醋栗或覆盆子等灌木作物，它们同样可以持续生长，如果是葡萄藤，则可以生长很多年。但农民种植的大部分作物都是一年生或（少数）二年生的。在温带气候下，这意味着作物通常在夏季生长，秋季收获。

农民在准备苗床时，有时先犁地并掩埋前一批作物的剩余物，继而开垦并打碎土块，然后用条播机成行种下种子。小麦或大麦等谷物可以在秋季播种，并在真正的寒冷天气出现之前就已经定植了。等到春季土壤变暖时，它们就可以生长了，这样可以最大限度地延长它们进行最有效的光合作用的时间。秋季可用于准备苗床的时间可能有限，或者冬季非常严寒，以至于许多种子作物的品种都要在春季播种。例如在加拿大北部，作物可能只有90天的时间完成从播种到收获的周期。马铃薯和甜菜等根茎类作物通常在春季播种，在当年秋季收获。一旦作物定植后,可施一次或多次肥料、除草剂和杀真菌剂。谷物和油料作物的收获时间是在谷物灌浆成熟后,马铃薯和甜菜的收获时间则是在预计作物几乎不会再继续生长时。

这显然是对这一过程的一个非常基本的概述,应该明确,它涉及许多不同的决策,准备苗床的最佳方式,种植哪些品种,采用怎样的播种率来产生最佳的植物种群,使用多少肥料以及何时施用,哪些杂草和病虫害可能会威胁到作物以及如何防治它们,何时收获,等等。不同农场、不同田地,甚至在同一田地内,所有这些都可能有所不同。

农民还必须决定种植哪种作物组合。有时,他们年复一年地种植同一种作物,但这可能导致病虫害的积累,后者必须加以控制,也可能导致土壤养分的消耗,而这些养分必须加以补充。在种植最有利可图的作物与保持作物健康和土壤肥力之间需要取得平衡。

在欧洲,现代的六茬耕地轮作的方式可能是先种两年小麦,再种一年菜豆,然后是小麦,再是大麦,最后是油菜(油菜籽)。菜豆打破了小麦的病虫害生命周期,并且作为豆科植物,有助于土壤固氮;油菜也打破了谷类病害的生命周期。经典的诺福克四茬轮作在小麦之后是大量喷洒农家肥、同时仔细除过草的芜菁,这样就为接下来的大麦作物提供了肥沃而无杂草的土壤,然后是白车轴草,白车轴草能够固氮,为接下来的小麦作物提供了肥沃的土壤,以此类推。现代化肥和农药使农民增加了补肥作物之间的时间,但这个原理仍然适用,而且比完全依靠化学药品便宜。轮作得到了简化,而不是淘汰了。

在美国中西部各州,也就是常被称为"玉米带"的地区,农民现在依靠的是玉米（在英国，这种作物被称作maize，而corn在那里是小麦、大麦和燕麦的统称）和大豆的轮作。在南方各州，玉米可能与棉花和一种豆类（如豇豆）轮作。在一些热带国家存在这一轮作主题的变体，那里像木薯这样能在地里生长数年的作物可以单独种植，也可以与一年生作物间种。在一些水稻种植区，土地可以一直休耕，直到播种下一季水稻作物为止，但在东南亚的某些地区，由于土壤适合在旱季种植，水稻可以与豆类、玉米或蔬菜轮作。

农民在决定轮作的最佳作物时，还必须考虑如何收割：前面提到的小麦/菜豆/大麦/油菜轮作中的所有作物都可以进行联合收割，而插入牧草则需要农场有放牧的动物，或者需要额外的机械来割草，并将其制成干草或青贮饲料。

因此，农民对其种植的作物所做的决定取决于一系列因素：不仅涉及土壤和气候允许他们种植什么，还包括他们拥有的知识和设备能够生产什么，以及他们或其动物需要吃些什么，或者能够卖什么。以下各章将讨论这些考虑因素。

第二章
农场动物

在现有的动物物种中,人类驯化的比例相对较小。正如美国博学家贾雷德·戴蒙德所言,动物需要有某些特征才适合驯化:合适的饮食,通常以多种植物为主,生长速度相对较快,有圈养繁殖的能力,以及合适的性情与社会组织,这样它们才不会惊慌失措,并且有完善的优势等级制度,以便人类可以接管它们成为领导者。只有少数物种结合了所有这些特征,成为最常见的农场动物。表2所示为十大(即最常饲养的)农场动物。

这些数字需要谨慎使用,因为一只鸡产生的食物量显然与一头牛、一只羊或任何更大体积的动物不可同日而语。它们还需要设置在具体的时代背景下。自1992年以来的二十年间,世界上鸡鸭的数量大约增加了一倍,山羊的数量增加了65%,而牛和猪的数量只增加了13%左右,绵羊的数量则没有变化。此外,农民还饲养了其他没有出现在这份名单上的动物,如南美洲的美洲驼、羊驼和豚鼠,牦牛,甚至还有蚕和蜜蜂等昆虫。

表2　2012年世界农场动物的数量（百万头或只）

鸡	21 867	猪	966
牛	1 485	火鸡	476
鸭	1 316	鹅与珠鸡	382
绵羊	1 169	水牛	199
山羊	996	骆驼	27

联合国粮食及农业组织统计资料库。《活体动物：头数》（联合国粮农组织，2015年）。2014年7月29日访问 <http://faostat3.fao.org/faostat-gateway/go/to/browse/Q/QA/E>。

对驯鹿、麝牛或鸵鸟等动物的最佳描述是半驯化动物，因为它们虽然被人们所利用，但没有像大多数农场牲畜那样受到集约化管理。马和驴等其他动物不仅用于农耕，还用于非农业运输，世界上某些地方的牛也是如此。因此，估量纯粹为农业目的而饲养的牲畜数量既困难，也不一定有用。

许多农场动物具有多种功能。牛产奶（在非洲的一些地方它们还产血），在世界的某些地方，牛在生命存续期间拉犁拖车，死后生产肉和皮；绵羊被剪掉羊毛，产奶，也被宰杀吃肉；骆驼和水牛既产奶又拉犁。

人类完全可能靠素食生活，彻底放弃动物产品，在发达国家，我们不再依靠动物纤维做衣服，也不再依靠畜力做牵引，那我们为什么还要养动物呢？难道仅仅是因为动物能生产出素食者不一定能轻易得到的方便的营养素集合？至少在部分程度上，这必定是传统的问题。在蒸汽机和内燃机发明之前，许多动物——马、驴、骡、牛，甚至狗——被饲养都是为了它们的肌肉力量，用来拖车拉犁，以及其他农具。它们的奶和肉是人们逐渐产生兴趣的副产品。这个论点显然不适用于猪和鸡,它们的价值是作为清道夫,通过吃食物残渣以及橡子或昆虫等大多数人类不能吃的东西来提供肉和蛋。因此,在数千年与家养动物一起生活的过程中,人类已经习惯了食用(并愿意花钱购买)它们生产的肉、蛋和奶制品,也找到了无数种方法将它们组合在一起,并与植物产品相结合,制成我们所吃的食物。因此,农场主们都想让他们的动物快速生长,并长成屠夫喜欢的肌肉和脂肪的正确组合,或者生产出大量的奶、蛋或毛,本章谈的就是他们为此需要了解些什么。

饲养农场动物

所有类型的畜牧业(即通常所说的农场动物照管)都普遍适用一些基本原则。归结到根本,并暂时忽略农场动物是有意识的生物,所有这一切就意味着畜牧业就是要把植物中的碳水化合物、蛋白质、脂肪等转化为动物中的碳水化合物、蛋白质、脂肪等,这样它们才能给我们提供肉和奶,在某些情况下,还有能量来拖动我们的农具。

动物拥有将植物中的淀粉分解为其组分葡萄糖分子所需的酶(生化催化剂)。葡萄糖直接用于为动物提供能量,或为了储存能量而合成为糖原。在乳腺中,它与其他分子结合形成乳糖,这是乳汁中主要的碳水化合物。尽管一些细菌和真菌具有将植物组织中的纤维素(膳食纤维)分解成葡萄糖所需的酶,但哺乳动物和鸟类并不具备这种酶。这就意味着,那些进化到能与含有合适细菌的肠道菌群一起生活的动物,可以从含有高比例纤维素的相对纤维性的叶子中提取营养。而那些没有进化到这一步的动物则必须吃种子和块茎,这些种子和块茎须含有它们能代谢的淀粉。

这是一个至关重要的区别,因为这意味着纤维素代谢者(如牛、绵羊、山羊、骆驼和水牛)不必与淀粉代谢者(如鸡、猪和人)竞争食物。当然,这两类动物日粮中的碳水化合物实际上包括糖类、淀粉和纤维素的混合物,但如何处理它们取决于各种动物的肠道解剖结构。

动物的肌肉、皮、毛发或绒毛,甚至部分骨骼,都是由蛋白质构成的。蛋白质是由较简单的化合物(被称为氨基酸,大约有20种)连接而成的大分子。植物和细菌可以合成全部20种氨基酸,但其中有10种是动物不能合成的。这些被称为必需氨基酸,它们必须在饮食中提供。简单地说,动物的消化代谢包括将食物中的蛋白质分解成它们的氨基酸组分,然后这些氨基酸便可用于形成动物本身的不同蛋白质。

动物体内的其他主要组分是各种脂肪,它们不仅沉积在皮下,而且存在于肌肉和肌纤维内部以及彼此之间,还构成了神经、细胞膜和激素的一部分。同样,这些大部分可以由动物利用食物中的油脂合成。还有一些从重量上来说属于次要的营养素,却是动物食物的重要组成部分。这些包括矿物质和维生素,缺乏它们会妨碍有效的新陈代谢,通常表现为疾病,如维生素D缺乏引起的骨骼变形(佝偻病)。

食物被吃掉后会发生什么取决于动物消化道的解剖结构。单胃动物在胃里开始食物的化学分解过程,然后半消化的物质进入小肠,对碳水化合物、蛋白质和脂类剩余的消化大部分发生在那里。剩余的物质则通过肌肉收缩进入大肠,在大肠中吸收水分,然后在排便过程中排出未消化的废物。实质上人类也是如此,如前所述,人类几乎没有消化纤维素的能力。

猪的消化道与此类似,只是盲肠(大肠头端的那部分肠道)比人的大,而且有微生物群,能有限地分解纤维素,所以猪最多可从这一来源中获得20%的能量。这意味着猪可以从牧草和其他植物纤维中获得一些食物价值,而这些纤维在人体内会直接穿肠而过。

鸡的消化道与单胃哺乳动物直至小肠的消化道在细节上完全不同,但从农户的角度来看,结果大同小异:它需要相对低纤维的饮食。然而,马和驴的盲肠体积要大得多,其微生物群具有相当大的分解纤维素的能力。因此,青草和干草等粗饲料可以构成马科动物日粮的很大一部分。

牛、绵羊、山羊和水牛(以及解剖结构不同的骆驼)等反刍动物也携带着可以分解纤维素的细菌,但都位于一个被称为瘤胃的大而复杂的胃中。反刍动物进食后可以让食物回到口中再次咀嚼——这就是反刍的过程,或称咀嚼反刍——这有助于分解纤维素食物,以便瘤胃细菌进一步代谢。事实上,如果没有一些膳食纤维,反刍动物的消化系统就不能正常工作,这意味着尽管例如高产牛可以得到含有淀粉、蛋白质和脂肪的精料,但如果要避免出现消化问题,它们也需要纤维物质。此外,对于奶牛来说,高精料/低牧草的日粮会降低牛奶中的脂肪含量。

从农户的角度来看,反刍动物和单胃动物之间的这些差异很重要,因为它们意味着反刍动物可以靠植物叶子的高纤维食物生活成长,而单胃动物则难以生存。因此,它们可以利用牧草,这在可耕地上是一种很好的中断作物(见第一章),还可以利用因过于陡峭、多岩或贫瘠而无法耕种的土地上的植被。

因此,农民饲养动物的方式取决于动物的种类及相应的解剖结构和生理功能、可用的饲料,以及饲养动物的目的。所有动物都有维持体重和各种生理功能所需的营养物质的"保养"需求,否则它们最终会饿死。大多数动物也有生长、怀孕、哺乳和活动的"生产"需求。

对于用来屠宰吃肉的动物来说,主要的生产成本通常是生长所需的饲养。在幼年动物中,首先达到生长速度峰值的是骨骼。随着动物年龄的增长,骨骼的相对生长速度下降,而肌肉的生长速度达到了顶峰。最后,随着肌肉生长开始减缓,脂肪沉积达到最大值(图5)。一般来说,在一个品种中,年龄较大或生长较快的动物脂肪含量较高。

农户试图在动物脂肪含量处于最佳状态时出售动物以供屠宰:脂肪含量太少,肉质很可能干柴无味;脂肪含量太多,动物就会失去价值。但何为"太多"会随着时间的推移而改变。例如,在19世纪,猪的体重往往比现在大得多,因为当时猪油或皮下脂肪的市场更大。

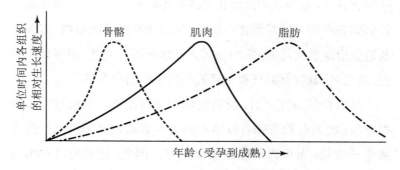

图5 动物从受孕到成熟时身体各组织的相对生长速度

动物怀孕时的生产需求明显增加，因为除了维持自己的身体外，它还必须应付正在成长的后代的需求。虽然这在后代出生后就停止了，但它还必须分泌乳汁来喂养它，这又需要额外的营养物质，超过了它本身的维持需求。实际上，奶牛的怀孕和泌乳可能同时发生，因为奶牛在产后两三个月就能再次受孕，而其泌乳期可持续九个月或十个月。在高产奶牛群中，农户会尽量让奶牛每年都生一头小牛，这样随着牛奶产量从高峰期下降，胚胎发育的需求就会增加。高产奶牛可能需要三倍或更多的维持量来满足这些生产需求。

对于生产大量牛奶的奶牛，单纯使用草料很难获得足够的营养物质，因此通常会给它们额外的高能和蛋白质精料日粮。即便如此，许多高产奶牛仍不能获得足够的营养物质满足其需要，并会利用背膘储备。另一方面，吃新鲜春季草的母羊，由于春季草易消化，蛋白质含量高，即使没有额外的精料饲喂，也可以喂养羔羊。

鸡的相应情况是产蛋。虽然大多数母鸡大概可以在农家院附近觅食，找到自己的维持口粮，但高产系统中的母鸡几乎每天都要下蛋，需要营养成分高的饲料。同样，整天拉犁的马需要的饲料是燕麦，或燕麦与豆类的混合，而如果不工作，需要的不过是青草而已。

农场动物的育种

除了喂养动物以获得最大的性能外，农民还可以通过动物育种的方式来影响其产量。区分品种和物种很重要。从本质上讲，物种是不能杂交的：通过杂交绵羊和袋鼠来繁育毛茸茸的跳跃动物的笑话在生物学上是不可能的。但在同一物种内，外观截然不同的动物可以成功地进行杂交。用公驴给母马授精就产生了骡子——几个世纪以来，骡子一直被用作驮兽和牵引畜力——印度瘤牛（*Bos indicus*）也可以与欧洲家牛（*Bos taurus*）成功杂交。举例来说，20世纪初，美国的育种者将瘤牛的品种"婆罗门"与原产于苏格兰的安格斯牛杂交，产生了肉牛品种"布兰格斯"。甚至还有人将欧洲家牛与北美水牛或野牛（*Bos bison*）杂交，形成了一个被称为"凯他洛"或"皮弗娄"的杂交品种。

因此，同一物种的各个品种之间的区别在于颜色、大小和/或形状上的差异，这些差异代代相传。这些差异可能源自地理上的差别，早期，各地农民生产的动物适合他们当地的特殊资源和需求。随着交流的日益增加，外地的农民意识到，他们可以使用其他地区的牲畜来改良自己的畜群。例如在18世纪，英格兰东北部的农民从荷兰进口公牛，这些公牛对于当地牛群的改良"功不可没"。20世纪初对英国牲畜的调查列出了13个主要牛种、8个猪种和29个羊种，此外还有各种杂交和小品种。从1960年代起，又从欧洲大陆引进了更多的牛品种，如"荷斯坦""夏洛莱""利木赞""西门塔尔"和其他几个品种。

不同的品种是为了不同的目的而培育的。如荷斯坦牛等专业的奶牛品种并不特别擅长产肉，而像阿伯丁安格斯牛这样的肉牛品种则不是为了生产大量的牛奶而培育的。但这些专门的肉牛开始在皮下和肌肉内覆盖脂肪时的体重要低于专门的乳牛品种，这样它们就可以"断草催肥"。换句话说，它们无需大量的精料饲喂就能达到屠宰体重。

美利奴羊经过培育能生产大量的细羊毛,但产奶量低,羔羊少,肉质差,而与萨福克或特克塞尔品种的公羊杂交的多产的骡羊或苏格兰混血母羊却奶量充沛,能培育出胴体质量高的双羔。美国的绵羊育种者将母羊区分为"朗布依埃""考力代""塔尔基"或"边境莱斯特"等品种,它们以其多产、体形、产奶能力和长寿而闻名,并分出了具有良好生长速度和胴体特性的"萨福克""切维厄特"或"特克塞尔"等公羊品种。

培育成能下大量鸡蛋的现代母鸡在产蛋期结束时,并不会作为可以直接烹调的鸡肉出现在超市货架上。它们是"淘汰母鸡",只能作为废物处置或用作宠物食品。骑乘马、小马以及骑乘骆驼的形状和大小与挽马和驮驼不同。所有这些数以千计的不同家养品种都是同一原始野生祖先的后裔,但经过人类数千年的选择,出现了形状、大小、属性各异的种类繁多的动物。

传统的育种者是靠眼睛和记忆来工作的。也就是说,他们根据自己希望产生的特性来选择要交配的公畜和母畜。他们希望体形超常的公牛与体形超常的母牛交配,产出快速生长的大型后代,或者高产奶牛与另一头高产奶牛的雄性后代交配,产出高产的雌性后代。他们的期望经常得到满足,从而培育出了现代农场中仍能见到的传统品种。

英国在19世纪末出现了育种协会,对各个动物的祖先进行记录,并举办展览,让育种者相互竞争,培育出最好的马、牛、羊、猪,以高价出售。对于育种而言,理想的动物是以颜色和形状来描述的。20世纪,这一过程被进一步推进,增加了牛奶产量的记录,这使育种者能够根据他们希望最大限度提高的性能来选择动物的配种。这是一种性能测试的形式,显然在产肉动物身上进行这种测试更为困难。因为胴体不能育种！然而，可以对动物及其食物进行称重，从而测量出它们的生长速度和饲料转化效率。在过去的四十多年里，人们已经开发出超声波扫描仪，可用于测量活体动物的背膘水平，从而评估它们的胴体成分。

测量动物体内没有表现出来的特性，如公畜所传播的基因对产奶量的影响等，可以通过测试后代来进行。例如，可以通过比较公牛雌性后代的产量和同一牛群中其他公牛雌性后代的产量，来评估公牛培育高产雌性后代的遗传能力，这一过程被称为同期比较。

动物的某些特征是由一对单基因控制的。例如，阿伯丁安格斯牛的黑毛色相比于赫里福德牛的红毛色永远是显性的，所以安格斯/赫里福德杂交牛总是黑身白脸。然而，大多数在经济上重要的动物属性都是由许多基因控制的，所以单一育种事件的结果不太容易预测。

因此，有三个因素对育种者能在多大程度上引发动物种群的遗传变化影响很大：遗传性、妊娠期和可用的动物数量。遗传性表达了个体与种群平均值之间可归因于遗传而非环境的任何差异的比例，分值为 0～1。高分意味着后代与亲本相似。例如，泌乳量的遗传性在 0.2～0.3 之间，而牛奶脂肪含量的遗传性在 0.5～0.6 之间。如此说来，改变奶牛群的牛奶平均脂肪含量就比增加其产量更容易。

妊娠期——后代在子宫内度过的时间长度（表3）——也很重要，因为繁殖新一代的速度越快，育种者引发遗传变化的速度也就越快。最后，可供育种者选择的动物数量越多，选择压力就越大。百里挑一的最佳动物很可能比十里挑一的更好。因此，我们预计家禽和猪的遗传改良会比牛的更快。这是因为一年内可以生产两代猪或家禽，而且一小块地方就可以饲养许多。两代牛之间可能会相隔三年之久，而且每公顷土地上只能饲养几头牛。

表3　大致的妊娠期

	平均天数	
骆驼	406	12～15个月
马	340	逾11个月
水牛	300～340	10～11个月
牛	283	9个月零10天
绵羊	150	大约5个月
山羊	156	大约5个月
猪	116	3个月3个星期零3天

　　猪和家禽的这种快速变化，在实践中也正是如此。从1947年到1988年，美国的家禽育种者将肉鸡养到准备屠宰所需的时间和饲料量减少了一半。现在一只两公斤的肉鸡可以在六周内生产出来。大致在同一时期，肉牛的生长速度也加快了，但只加速了大约16%。现在，大多数商品猪和家禽都是由专门的育种公司雇用专业遗传学家培育出来的。

　　牛的育种还没有实现这种程度的专业化，但人工授精（AI）特别有助于奶农获得高质量的遗传物质。这意味着他们不必再饲养自己的公牛，而是可以使用经过育种性能测试的公牛精液

（以冷冻的形式到场）。由于人工授精使用的是在高选择压力下挑出来的公牛，因此遗传改良比自然交配更迅速。此外，人工授精还可以将优秀的公牛用于更多的母牛。一些专业育种者还采用胚胎移植技术，将优质母牛的多个胚胎从原牛身上冲出，分别移植到受体动物身上，从而使性能较差的母牛获得优质的后代。

安置农场动物

除了喂养和繁殖动物外，农民还试图通过安置动物来提高其生产力。农场动物一生基本上都能够在户外生活，虽然母鸡喜欢在窝里下蛋。一些住在户外的母猪如果能找到足够的合适材料，也会筑起大窝，在里面生下小猪。但是，养殖户在喂养了动物之后，往往不愿意看到口粮中的部分能量成分在寒冷、刮风、下雨的时期被用于保暖。这对出生在户外的小羊来说是一个特别致命的组合，尽管它们完全能够在一个晴朗、平静而寒冷的夜晚生存下来。相反，在炎热的气候下，动物可能需要保护免受阳光的照射，而水牛需要在酷热的时候找到水源，沉浸其中。

农场提供的住宿条件千差万别。一个极端是一些简单的建筑，不过是用来挡雨的屋顶；而在另一个极端，有的房舍具备人工照明、通风和温度控制系统，育肥猪和肉鸡可以在其中度过一生。

从某种意义上说，这种水平的安置对农民和动物都有好处。与散养的鸡群相比，蛋鸡舍更容易实现鸡蛋采集和喂养的自动化，从而降低了劳动成本。此外，母鸡产蛋会因日长增加而受到激励，因此如果这些禽类依赖人工照明，农民就可以更好地控制其产蛋模式。即使是简单的牛棚也意味着农民可以决定它们所产生的农家肥的用途，事实上，这也是对育肥牛进行安置的早期论点之一。不过毫无疑问,安置会影响动物的行为,特别是对其一生的集约式安置,这导致了有关动物福利的争论,第五章将对此进行详细讨论。

安置对健康也有影响。虽然农民为保护动物不受天气影响而安置它们,但其结果可能是让它们面临更大的疾病风险。例如,通常用药物来治疗安置的家禽,以抑制各种导致球虫病(一种肠道感染)的艾美耳球虫(*Eimeria*)的发展。拥挤的房舍可能也会像饥渴那样引起动物的应激,而应激的动物更容易患病。

农场动物的健康与疾病

农场动物和人类一样,可能会患上各种各样的疾病,而且它们无法告诉农场主自己不舒服,这更增加了难度。畜牧业优秀管理者的特征之一是能够发现疾病的早期迹象,以便隔离和/或治疗动物,并了解可能会导致动物出现问题的情况。

膘肥体壮的高产奶牛有时会在产犊几天后毫无征兆地倒下,这种疾病被称为"产乳热",如果不加治疗,可能会致命。其原因是血钙水平下降,只要农民知道怎么做,注射硼葡萄糖酸钙就能轻松治愈。这只是各种矿物质或维生素短缺或过剩可能造成的问题的一个例子。

动物还会受到蝇、虱子和蜱虫等体外寄生虫的攻击。例如,在美国东南部各州,新世界螺旋蝇(*Cochliomyia hominivorax*)的蛆虫会钻进牛的肉里。一个取得了成功的控制计划是在交配季节的高峰期释放实验室培育的绝育雄蝇,这样在几代的时间里就减少了蝇群的数量,到1980年代,这种蝇已经被灭绝了。

在非洲控制采采蝇(舌蝇属的各物种)的尝试则不太成功,

仍有大片地区的牛在一年中的部分时间里或全年都不能放牧。问题并不在于蝇本身,而在于它有能力用锥虫感染被它叮咬的动物。锥虫是一种生活在血液中的小型鞭毛虫生物体,会引起人的昏睡病和相关的疾病(疾病的一个名称是"那加那"),包括贫血、发烧和牛的缓慢渐进性消瘦。

同样,种类繁多的蜱虫也可能是各种疾病的携带者,如非洲牛的血尿热、东海岸热、胆管热、胆囊病等,因此需要每周喷药防治蜱虫。此外还有肝吸虫和线虫等体内寄生虫,引起破伤风的梭菌属细菌,导致牛瘟和口蹄疫的病毒等。所有这些都是养殖的潜在问题来源。

过去这半个多世纪,兽医业和制药业研制了许多产品来帮助农民防治这些疾病。有防治线虫的驱虫药,有防治梭菌属细菌的疫苗,还有防治各种细菌感染的抗生素。个体农民需要考虑的问题是,治疗的成本是大于还是小于疾病造成的利润损失。在某些情况下这似乎没有什么疑问,他们做出的决定是:与其承担预防药物的成本,不如接受其畜禽群的一些死亡风险。

然而,对于诸如口蹄疫等某些疾病,国家往往认为疾病的潜在影响非常大,必须对其采取措施,甚至包括宰杀受感染的动物,尤其是在"人畜共患病"的情况下,这种疾病有可能会从动物传染给人。在过去的几年里,英国出现了许多这样的例子:牛海绵状脑病(BSE)、高致病性禽流感(HPAI)、大肠杆菌O157、沙门氏菌、弯曲杆菌、李斯特菌和牛结核病(bTB)。

动物生产体系

考虑到所有这些饲养、育种、安置和疾病控制的因素,农民就必须对生产体系做出决策。有多少家不同的农户,就差不多有多少种饲养动物的方式,但大多数可以根据密集度、专业化和规模来分类。集约化体系是指将动物集中在面积相对较小的土地上,但使用大量的劳动力、设备、房舍和精饲料,而广泛化体系则使用更多的土地和较少的其他投入。因此,发达国家的大多数商业性养猪和养禽单位都被称为集约型,而东非四处游牧的牧牛人或美国达科他州的大牧场主运营的则是广泛化体系。

一些专门的畜牧农场可能只有猪或家禽,而在混合农场中,可能会经营几种动物和作物,如传统的欧洲混合农场饲有奶牛和肉牛、绵羊、少量猪和家禽,全部由农场生产的牧草、根茎作物和谷物喂养。与此相反,专门的西欧乳品农场则只饲养奶牛,也许还有一些幼牛作为牛群的替代品,大部分土地用于种植青草,并种植一些玉米作为青贮饲料。

从只有几头猪几只鸡的零散饲养到在数千公顷土地上饲养数千头牲畜的大型单位,畜牧业的规模几乎是无限变化的。然而,无论其经营规模、密集度或专业化的程度如何,大多数畜牧业者都有一些相同的工作要做。

所有的畜牧业者都需要确保动物每天都能获得食物和水。对于一些人来说,这很容易,因为他们的动物在田野或牧场上放牧,他们甚至可能不会每天查看畜群;或者只需按一下按钮,自动喂食器就会开始工作。但对其他的人来说,特别是在温带气候的冬季,或在热带的干燥季节,这可能会占用大量时间,需要他们移动大量的干草、青贮和精饲料。

如果他们有产奶的动物,还需要保证每天给它们挤奶。一头奶牛在产犊后,会产奶长达九或十个月(山羊的产奶期可能会长得多，有的甚至会在没有生过羊羔的情况下就产奶），最简单的方法就是让牛犊来吸取乳房里的奶水。一些饲养所谓的喂犊母牛的肉牛养殖者就是这样操作的，这是牛犊生命的良好开端，但显然不能生产供人食用或出售的牛奶。而且牛犊所需要的奶量与现代西欧的专业奶牛的产奶量完全不同，后者在305天哺乳期内的平均奶量超过7 000升。

因此，几千年来，农民一直在限制牛犊接触母牛，并至少为自己挤走部分牛奶。其间大多数时候都是用手来完成的，他们在早上和傍晚时分坐在奶牛身边，用手部动作来模仿牛犊的吸吮。直到20世纪才发明了有成效的挤奶机器，而一旦有了挤奶机，农民就可以饲养更多的奶牛，也就不需要再大批雇用挤奶工了。近年来开发了全自动挤奶机，这意味着奶牛可以自行选择何时挤奶。许多奶牛选择每天挤奶两次以上。

畜牧业者需要了解动物的正常行为。奶牛愿意接受牛群中其他奶牛的骑跨，就表示它们已为交配做好了准备。如果它们和公牛厮混在一处，这是养殖户最感兴趣的，但对他来说对奶牛进行人工授精的时间是很重要的，因为在正确的时间即奶牛"发情"时进行人工授精，决定了它是否会受孕。对人类来说，绵羊的发情检测要困难得多，只能让公羊来完成。

了解动物何时分娩也很重要。奶牛经常会离开牛群走到田间一角，在户外饲养的母猪会收集稻草和树枝，就像要搭窝一样。在大型畜群中，并不总能做到这种个别观测，可以根据孕畜交配后的时间长短将其移入产犊（"farrowing"是英文中指代怀孕母猪的术语）的房舍。异常行为也可能是动物患病的信号，农民的部分工作就是进行常规注射，为动物注射疫苗或杀死它们的肠道寄生虫，或者检查羊蹄，必要时修蹄并进行药物治疗，以防止其跛行。

清洁是疾病控制的一个重要部分，农民往往会花费大量时间运粪，或在两批动物的间隙清理畜舍，以避免感染性微生物在不同批次之间的转移。出于同样的原因，他们也会将牲畜转移到干净的草场上，并确保它们待在合适的田地里，这意味着必须花时间维护栅栏和篱笆。

农民还需要知道运往肉类市场的动物何时可以屠宰，这可能只不过是检查它们是否达到了屠宰场规定的重量，但也可能需要沿着背部皮肤触摸骨头来评估皮下脂肪的覆盖情况。

所有的任务都完成后，放牧牲畜的农民还必须确保他们在生长季节保存了足够的饲料，让牲畜能够度过冬季或旱季。

虽然许多现代动物可能永远见不到绿地了，特别是猪和鸡，越来越多的奶牛也是如此，但它们的基本生物学特征仍然决定了农民利用它们生产食物和纤维的方式。因此，农民永远不能忘记它们也是活的生物体，而不仅仅是商品。

第三章
农产品与贸易

正如我们在第一章和第二章中所看到的那样,农民花费了大量时间种植作物,饲养牲畜,通常超过了其家庭的需要。他们的部分时间必须用来出售剩余的产品。本章介绍的是整个农业产业的生产情况,以及出售其产出的当地和国际市场的运作情况。

很明显,农民为自己和其他人生产食物,但他们也为农场动物(以及伴侣动物——宠物)生产饲料。这是最大的两类农产品,但远不止这些:植物纤维,如棉花或黄麻;动物纤维,主要是羊毛;从沼气到干骆驼粪等各种形式的燃料;以及药品,如从猪胰腺中提取的胰岛素,还有从猪脑中提取的用于治疗阿尔茨海默病的脑活素。清单在哪里停止,取决于如何界定我们所谓的农业。它当然包括烟草种植,但罂粟的种植呢?我们也许认为花卉是由园丁种植的,或者在商业规模上,是由我们称之为"市场园丁"的人在小块土地上种植的,但越来越多的花卉是以田野的规模种植的,任何开车路过荷兰部分地区的人都会看到;鉴于那里的某些温室如此庞大,我们几乎可以说那是工厂的规模了。这同样适用于多种沙拉作物,如番茄、甜椒和莴苣。那菠萝、茶或橡胶树等种植园作物呢?或者诸如咖啡或可可等产自小树或灌木的作物?

同样有必要记住,农民不仅仅生产物质产品。他们负责提供的服务通常被归类为"生态系统服务":农田是许多不同种类的野生动物的栖息地;农民可以使用(或允许其他人使用)他们的土地作为太阳能电池板或风力发电的场地;许多农业景观赏心悦目,因此,农业也为旅游业提供了资源。

与粮食生产相比,燃料、药品、生态系统服务等的相对重要性难以确定,在世界范围内必然如此。但毫无疑问,粮食是农业主要和最重要的产出。农场规模越小,农业经济越不发达的地方越是这样。蔬菜、沙拉作物、水果、坚果、马铃薯和其他根茎类作物、牛奶、鸡蛋和鸡都可以种植和加工到可以食用的阶段,而不需要任何超出农户家庭通常可利用的设备或专门知识。

然而,如表4所示,世界上绝大部分农业产出是以产品的形式出现的,这些产品需要更多的加工,超出了大多数家庭现有的能力。此外,由于如今的城镇人口比农村人口多,绝大多数人无法自己种植粮食。因此,只要农民生产的粮食超过了家人的消耗量,他们就应该被看作食品工业的原料供应商,而且经营规模越大,就越有可能出现这种情况。即使是一离开土地就可以进厨房的马铃薯和洋葱,或者从树上摘下来就可以吃的苹果,也需要储存起来,以延长它们的供应季节,并运输到消费者可以购买的地方。

表4 主要作物：全球种植面积、产出及平均产量

	面积 （百万公顷）	产出 （百万吨）	平均产量 （吨/公顷）	干物质产量 （吨/公顷）
谷物				
小麦	215	670	3.1	2.7
玉米	177	872	4.9	4.3
水稻	163	720	4.4	3.9
大麦	50	133	2.7	—
高粱	38	57	1.4	1.1
谷子	32	30	0.9	0.7
燕麦	10	21	2.2	—
黑麦	6	15	2.6	—
糖料作物				
甘蔗	26	1 832	70.2	7.9
甜菜	5	270	55.0	—
根茎类作物				
木薯	20	263	12.9	4.9
马铃薯	19	365	18.9	3.8
番薯	8	103	12.8	3.8
薯蓣	5	59	11.7	3.5
油料作物				
大豆	105	242	2.3	—
油菜籽	34	65	1.9	—

续表

	面积 （百万公顷）	产出 （百万吨）	平均产量 （吨/公顷）	干物质产量 （吨/公顷）
葵花籽	25	37	1.5	—
油棕榈果	17	250	14.5	—
橄榄	10	16	1.6	—
豆类				
菜豆（干）	29	24	0.8	—
鹰嘴豆	12	11	0.9	—
豇豆	11	6	0.5	—
兵豆	4	5	1.1	—
果蔬				
葡萄	7	67	9.6	—
香蕉	5	102	20.6	6.2
苹果	5	76	15.8	—
番茄	5	162	33.7	—
洋葱	4	83	19.7	—
饮料				
咖啡（生）	10	9	0.8	—
可可豆	10	5	0.5	—
茶	3	5	1.5	—

联合国粮食及农业组织统计资料库。产量（联合国粮农组织，2015年）。2014年4月26—27日访问 <http://faostat3.fao.org/browse/Q/QC/E>。

注：面积和产出的数字均已四舍五入，但平均产量的数字是原始数据。关于各国的种植面积，英国为2 270万公顷，法国5 500万公顷，肯尼亚5 800万公顷，墨西哥1.98亿公顷，以供比较。

经济越是工业化，这个看似简单的过程就变得越复杂。苹果会被分类、分级，也许还会贴上小标签，告诉你它们的品种。然后，它们可能需要被储存起来，也许要存上几个月，再被送到中央配送点，最后运到各个商店或超市。

在相反的极端，整个产业已经发展到加工谷物、牛奶和肉类的程度。农民将小麦卖给磨坊主，磨坊主将小麦磨成面粉（以及各种等级的麸皮，后者可以筛出作为动物饲料出售），然后卖给面包师。如果小麦的品种或质量不适合磨成面包粉或饼干粉，则会卖给动物饲料制造商。奶从奶牛或山羊身上挤出后可以直接饮用，但对于发达经济体的人来说，奶会在农场冷藏，从农场运到乳制品厂，经过巴氏杀菌或消毒，然后灌入瓶子或塑料容器。并非所有的奶制品都是流通市场所需要的，因此它们将被转化为黄油、酸奶或数百种（乃至数千种）奶酪中的一种。

想象一个农民试图将一头重达600公斤、出产330公斤带骨牛肉的育肥牛转化为可在烤箱或煎锅中烹调的牛肉时，会遇到哪些问题。即使能高效而人道地宰杀牛，也会有这样的问题：如何处理胴体中较难食用的部分，如何防止肉在食用前腐烂，又如何将胴体的各部位分离成最适合不同烹饪方式的大块带骨肉。这就是屠宰业数百年来一直是一个与农业相分离的技术性行业的原因，也是人们为何开发出了从传统的风干、盐渍、烟熏、腌制，到近代的冷藏、冷冻，甚至辐射处理等一系列食品保鲜方法的原因所在。

因此，农业需要被视为食品链的一部分，许多农民生产范围相对较小的基础产品，然后它们被转化为我们在家庭厨房或由食品加工和分销行业摆上餐桌的大量菜肴。表4列出了播种面积和产出最大的作物,表5列出了主要的动物产品。

请注意,表4中给出的产出和产量数字都是生鲜重量。这对谷物和豆类来说没有多大区别,但对根茎类来说却很重要。储存的稻谷和小麦含有约12%的水分和88%的干物质,而马铃薯只有大约20%的干物质。以世界平均产量数字为例,表4的右栏显示了这些差异对一些重要作物的影响。当然,有些农民的产量会超过世界平均水平:大多数欧洲农民希望他们的小麦产量是世界平均产量的两倍,否则他们是不会满意的,2010年新西

表5　2012年主要畜产品全世界总产量(千吨)

肉		奶	
猪	108 507	牛	626 184
鸡	92 731	水牛	97 942
牛	62 737	山羊	18 002
绵羊	8 481	绵羊	10 010
火鸡	5 634	骆驼	3 001
山羊	5 294		
鸭	4 358	**蛋**	
水牛	3 594	母鸡	66 293
鹅	2 798		
马	766	**羊毛**	2 067
骆驼	511		

联合国粮食及农业组织统计资料库。主要畜产品(联合国粮农组织,2015年)。2014年10月9日访问<http://faostat3.fao.org/faostat-gateway/go/to/browse/Q/QL/E>。

兰创造的小麦产量世界纪录是每公顷15.7吨。

表4只显示了一些主要作物。例如，它没有显示牧场的面积，根据一项计算，全世界的牧场面积为35亿公顷，其中包括专门用于苜蓿等饲料作物的3 500万公顷。准确的数字是一个定义问题，因为农民的动物与各种野生动物共享的大面积粗放型草场、稀树草原等并未计算在内。表4集中介绍了为人类和动物食物而种植的作物，但也有大约3 500万公顷的纤维作物，其中90%以上专门用来种棉花。

还必须记住，某些作物有多种用途。玉米就是一个很好的例子。我们熟悉的玉米穗轴，新鲜时作为甜玉米出售，或是做成玉米粒罐头。蒸熟碾碎后可制成压片玉米用作牲畜饲料，加入各种调味料就成了早餐时食用的脆玉米片。它们还能生产淀粉、油和玉米糖浆。包括穗轴和秸秆在内的整个作物都可以收获、切碎，并制成青贮饲料，现在是美国和欧洲牛群的主要冬季饲料之一。最近，为了应对油价上涨和增加可再生资源利用的立法要求，美国越来越多的玉米被转化为乙醇，加入汽油中。到2010年，用于制造乙醇的玉米比喂养牲畜的还多。

大豆的情况与之类似，它含有高达20%的油和40%的蛋白质。大豆被压碎并从其剩余部分中分离出油后，作为豆粕被用作牲畜饲料。豆油是美国使用最广泛的食用油（通常被标为"植物油"），是人造黄油、蛋黄酱和其他沙拉酱、各种烘焙产品，以及其他众多用途的基础，但到2011年，美国生产的所有豆油中有近四分之一被转化为生物柴油。其他作物的情况也如出一辙；农民不仅是食品链的一部分，也是其他许多产品链的一部分。

畜牧业者也使用了范围相对较窄的动物（表2）来生产广泛

得多的食品和其他产品。表5列出了占生产量比例最大的动物。与表2相比，我们可以看到，鸡的数量远远多于其他任何畜禽，却并没有转换成更大量的肉类产品，就是因为鸡的体积太小。

也许更令人惊奇的是，虽然世界上牛的数量始终都比猪多，而且每头牛胴体的产肉量是猪的两三倍，但猪肉（包括鲜猪肉、熏肉、火腿和各种形式的腌制猪肉）的产量却几乎是牛肉和小牛肉产量的两倍，因为猪长到待宰成猪的速度（大约6个月）比牛快，谷饲的小型牛出栏至少需要12个月，而草饲的牛则普遍需要18～24个月的时间。当然，很多牛群是奶牛，在第五年之前不太可能被屠宰，存续的时间可能会长得多。

中国是猪肉、鸡肉、绵羊肉和山羊肉的最大生产国，美国是牛肉和火鸡肉的最大生产国，印度和巴基斯坦主导着水牛肉的产量。骆驼肉是非洲国家唯一占据市场的产品，苏丹、索马里、肯尼亚和埃及都是较大的生产国。

1992—2012年的二十年间，许多这类动物产品的世界总产量大幅增加：鸡肉翻了一番，鸡蛋几乎翻了一番，山羊肉的产量增加了约85%，猪肉的产量增加了约50%。尽管有这些增长，但仍有大约40亿人以水稻、玉米或小麦作为主食。另有5亿人以木薯为食，木薯是一种起源于南美洲的块根作物，但如今在撒哈拉以南非洲被广泛种植和食用。它具有抗旱性，因此常被作为"饥荒作物"种植并留在地里，在其他主食稀缺时进行收割。在马里、尼日尔、孟加拉国和柬埔寨等非洲和亚洲的一些国家，谷物提供了70%以上的膳食能量，而在北美和西欧，这一数字还不到30%。

在这些较富裕的国家，大多数人的饮食以肉类和奶制品中的蛋白质和脂肪来补充谷物和马铃薯提供的碳水化合物。但生产这些产品的动物当然也需要饲喂，它们的饲料有的来自人们无法消化的牧草和其他高纤维植物，有的则来自谷类以及大豆等植物蛋白生产作物。2005年，加拿大种植的谷物中有70%以上用于饲喂动物，英国种植的小麦中有40%用于饲喂动物。在世界范围内，据估计，约有三分之一的谷物被用来饲喂动物。1993—2013年，阿根廷和巴西的合计大豆产量几乎翻了两番，因为那里的种植扩大到以前用于放牧或被森林覆盖的地区，世界上约80%的大豆收成都用于动物饲料。

这是否意味着阿根廷和巴西大幅增加了它们生产的动物数量，并因此消费了更多的肉类和更多的乳制品？不一定，因为正如我们前面看到的，必须把农业看作食品工业原料的生产者，或者说是食品链的第一环节。因此，任何农产品都可能在农场、当地市场上消费，也许是在生产该产品的国家，或者是在世界的其他地方。为了理解为什么会发生这种情况，我们需要更多地了解农业市场如何运作，这正是我们在本章下一节考察的问题。

农产品市场

英格兰东南部的肯特郡历来以其果场的面积和质量而闻名。2014年秋天，那里的苹果种植者期待着丰收。前一个冬天很冷，消灭了害虫，八月有足够的雨水帮助果实生长，九月又阳光明媚，采摘工作得以提前开始。欧洲其他地区的情况也大致相同。根据报纸的报道，一位农民不记得有哪一年比那年的长势更好，另一位农民说"看到收成喜人，精神相当振奋"。但他们都觉得自己的苹果要跌价，一个人做好了亏损的准备，另一个人觉得可能只够支付成本。这看似毫无道理，却并不是一个新问题。莎士比亚《麦克白》中的搬运工想象着自己为"眼看碰上了丰收的年头,就此上了吊的农民"打开了地狱之门。那么,令人高兴的收成怎么会导致经济损失呢？如果了解农产品市场的运作方式,应该就能解释这个问题。

与其他任何市场一样,农产品市场也受到需求和供应变化的影响。如果需求增加而供应量不变,价格就可能上升；如果供应量增加而需求量不变,价格就可能下降。因此,我们需要知道是什么影响了农产品的需求和供应。对一种产品的需求受其价格、潜在消费者的收入和口味,以及这些消费者的数量的影响。价格高的时候,消费者在购买产品之前会仔细考虑,价格低的时候,他们会考虑购买更多的产品。这是基本思路,当然也要看是什么产品。

对于水稻、玉米粉、意大利面、面包或马铃薯等主食,需求不会有太大的变化。价格高的时候,人们仍然需要基本的饮食,所以他们会付出自己必须支付的费用。而价格低的时候,他们不可能大量增加主食的消费量,而是会把省下来的钱用在其他产品上。相反,奢侈食品的市场受价格水平的影响会大得多,因为它们不是饮食的重要组成部分,所以价格的小幅上涨可能会导致消费数量的大幅减少,反之亦然。可替代性在这里也有影响。我们可以从价格较高的分割肉中获取蛋白质,这些肉的肉质细嫩,口感好,但也可以从便宜的肉中获取蛋白质。同样,只要消费者认为烤猪肉是烤牛肉的可接受替代品,牛肉的高价格就可能会增加对猪肉的需求。不过,有些消费者可能因为文化或宗教原因而拒绝吃任何形式的猪肉,因此他们的喜好不会影响猪肉市场。

这只是消费者口味影响需求的众多例子之一。许多工业化国家的政府出于健康原因，试图影响消费者的食品偏好，例如鼓励少吃红肉，多吃水果蔬菜。这些运动是否像商业广告一样成功是一个有争议的问题（这在通识读本《食物》和《营养》中都有详细讨论），商业广告是影响消费者口味的另一个主要因素。

收入水平以各种方式影响了对食品的需求。21世纪初，为美国农业部进行的一项研究比较了低收入国家和高收入国家消费者的行为，前者的收入不到美国的15%，后者的收入为美国水平的50%或以上。表6给出了一些研究结果。这些结果表明，富裕国家的人民在食品上的支出占其收入的比例比贫穷国家的要小，他们的食品消费行为受收入变化的影响也较小。随着人

表6　收入对食品需求的影响

	低收入国家	高收入国家
预算中用于食品的百分比	47	13
收入增加1%时，食品的需求量增加	0.73%	0.29%
食品预算中用于谷物的百分比	28	16
收入增加1%时，谷物消费量增加	0.56%	0.19%
食品预算中用于肉类的百分比	18	25
收入增加1%时，肉类消费量增加	0.82%	0.33%
食品预算中用于奶制品的百分比	9	14
收入增加1%时，奶制品消费量增加	0.93%	0.35%

A. 雷格米，M. S. 迪帕克，J. L. 小希尔，J. 伯恩斯坦，《食品消费模式的跨国分析》，美国农业部经济研究局，表B-1和表B-2可在 <http://www.ers.usda.gov/media/293593/wrs011d_1_.pdf> 处查阅（2014年10月20日访问）。

们越来越富裕，他们在蛋白质和脂肪食品上的花费会更多，比如肉类和奶制品，但对谷物等主食的消费增长速度较慢。该研究还发现，穷人比富人对食品价格变化的反应更灵敏，这也许并不奇怪。

这些数据对富裕国家的农民来说是个坏消息，因为它们意味着除非人口增加，否则对食品的需求不会有太大的增长，而富裕国家的人口增长往往低于贫穷国家。反过来，这些数据对贫穷国家的消费者来说也是个坏消息，因为它们意味着食品价格的小幅上涨或收入的小幅下降将对他们能够购买的食品数量产生巨大的影响。

需求增长缓慢对富裕国家的农民有影响吗？如果他们增加的供应量多于需求量，就会有影响，因为他们得到的价格就会下降。从理论上讲，这应该是一个自我调节的系统，价格下降会导致市场供应量减少，因为所有农民都会降低产量，或者有些农民完全破产，供应量因此下降到与需求量相当的水平。

实际情况更为复杂。产品价格不是影响农场产出的唯一因素。生产成本也会发生变化：例如，猪和家禽的利润率对饲料价格的变化非常敏感。农民可以采用新的作物或牲畜品种，在不需要更多投入的前提下生产更多的产品，或者采用其他各种技术变革中的任何一种来增加产出或降低成本。政府可以像第二次世界大战后及其后许多年的欧洲一样改变政策鼓励农民增产，或者当这种政策成功到产生盈余时，鼓励减产。

农业生产不是轻轻一按开关就能打开和关闭的，记住这一点也很重要。如果农民决定要生产更多的牛奶，那么他们可能需要两年多的时间才能将小母牛饲育到可以配种、怀孕、产犊的程度,并在奶牛群中占有一席之地。木本作物也可能需要数年时间才能投入生产。因此,总的来说,除了一些例外,许多农产品的产量往往对市场条件的变化反应缓慢。

肯特郡的苹果种植者对丰收喜忧参半的原因现在应该很清楚了。随着超市销售的水果种类增多,英国对苹果的需求量正在下降,价格下降不会对消费者购买苹果的数量产生太大的影响。此外,英国农民还要与其他国家的供应商竞争,他们的苹果也销往英国市场,这就涉及农民的另一个重要问题:国际贸易问题。

农产品的国际贸易

正如我们在前文中看到的那样,一些农民只生产他们和家人想吃的东西,而不出售他们的任何产出。他们是勉强维持生活的农民,而且人数不多。大多数农民大概都会出售他们的一些产品,就是因为他们需要从盐和铁到化肥和农药等自己无法生产的东西。农民几乎从农业开始时就会把粮食和纤维卖给非农民。这最终导致了区域专业化,因为良好耕地上的农民意识到,他们可以通过专门从事农作物生产来赚取更多的钱,而那些贫瘠的高地农场的农民则意识到,用牲畜换取他们所需的农作物,他们会过上更好的生活。这一进程的逻辑结果是,专门从事农业的国家应该出口农产品,以换取工业化国家生产的制成品。

这就是19世纪末出现的模式。当时的大部分国际贸易都是温带产品,那些产品从美国、加拿大、阿根廷、澳大利亚和新西兰等欧洲移民定居的国家运回欧洲,或从东欧和俄罗斯运去西欧。茶叶、咖啡、可可和含油种子等热带产品在总量中所占的比例很小。

表7　2012年十大农产品进口国和出口国

出口国	占世界出口量百分比	进口国	占世界进口量百分比
欧盟27国	37.0	欧盟27国	35.7
（欧盟国家向非欧盟国家出口）	（9.8）	（欧盟国家从非欧盟国家进口）	（9.9）
美国	10.4	中国	9.0
巴西	5.2	美国	8.1
中国	4.0	日本	5.4
加拿大	3.8	俄罗斯	2.4
印度尼西亚	2.7	加拿大	2.2
阿根廷	2.6	韩国	1.9
印度	2.6	沙特阿拉伯	1.7
泰国	2.5	墨西哥	1.6
澳大利亚	2.3	印度	1.5
十大出口国的出口量占世界总出口量的百分比（按价值计算）	73.1	十大进口国的进口量占世界总进口量的百分比（按价值计算）	69.5

世界贸易组织，《商品贸易，2013年国际贸易统计第二部分》，© 世界贸易组织，2013年，2014年10月访问 <http://www.wto.org/english/res_e/statis_e/its2013_e/its13_merch_trade_product_e.pdf, table II.15>。

当前的世界农产品贸易格局中仍然存在着这种情况，这一点或许出人意料。如表7所示，尽管一些新兴的工业化国家现在也加入了主要进口国的行列，但世界农业贸易仍然由工业化国家所主导。欧洲仍然是主要进口国，美国、加拿大、阿根廷和澳大利亚仍然跻身于十大出口国之列。欧盟、美国和中国同时出现在进口国和出口国的名单上，只是因为它们是某些商品的主要进口国和其他商品的主要出口国。例如，美国是小麦、水稻、玉米和牛肉的五大出口国之一，也是糖、咖啡、可可和茶的五大进口国之一。欧盟国家之间的贸易额相当大，这也是欧盟在这两个榜单中排名第一的原因，某些欧盟国家的贸易商既是进口商又是出口商：例如，德国是黄油和奶酪的主要进口国和出口国。

虽然大部分热带饮料——茶、咖啡和可可——是为出口而生产的，而且相当一部分大豆也是供出口的，但如表8所示，食品贸易一般只占世界产量的一小部分，往往不到20%。收成不好

表8 2011—2012年世界出口占世界产量的百分比

小麦	22	猪肉	13
玉米	13	鸡肉	13
水稻	5	牛肉	16
糖	22	绵羊肉	10
马铃薯	3		
香蕉	18	奶（等价物）	14
大豆	38		
咖啡	84	蛋	3
可可	64		
茶	40		

联合国粮食及农业组织统计资料库。农作物与畜产品。某些国家的出口数字（联合国粮农组织，2015年）。2014年10月10日访问 <http://faostat3.fao.org/browse/T/TP/E>。

时,通常首先满足国内需求,所以可供出口的较少,而丰收后的大部分盈余将被送往出口市场。因此,世界市场的供应量往往比世界产量波动更大,世界市场的价格也因而变化不定。

例如,想想一个假设的国家在丰收一亿吨水稻后,以世界平均百分比出口的情况。在正常年份,它将向世界市场供应500万吨。在产量增加了5%的好年景,它将有1 000万吨可供销售,而在产量下降5%的歉收年份,如果要满足其国内需求,则根本没有可供出口的余量。

各国政府对这些国际贸易和国内供求问题的关注是可以理解的。即使在供应多样而充足的工业化国家,粮食价格也可能是一个敏感的政治问题;在贫穷国家,供需平衡就可能是生死攸关的大事。因此,在大多数国家,政府都会对农业有某种参与,包括从统计监测投入和产出,或向农民提供咨询,到全面控制生产、价格和农业收入等。即使是那些赞成市场力量运作的工业化国家的政府,如欧盟或美国政府,也采取了支持农业收入和影响农产品贸易的措施。

在国际一级,世界贸易组织监测农业贸易,并试图通过定期召开政府间会议来减少自由贸易的壁垒。联合国粮食及农业组织收集全面的农业统计资料,并就世界农业状况和解决农业问题,特别是发展中国家的农业问题,编写了许多报告。

本章的目的是要表明,尽管农业的产出种类繁多,但只是为了确保人们每天有饭吃而出现的地方、区域、国家和国际复杂联系的一部分。这些联系还延伸到农业的另一面,以提供农民所需的投入,它们是第四章讨论的主题。

第四章

农业投入

　　有个故事，说一位老农接手了一块撂荒多年的土地。这块土地杂草丛生，大门挂在铰链上，树篱要么过于繁茂，要么满是缝隙，排水沟渠淤塞，低处的田地泥泞不堪。经过几年的辛勤劳动和大量投资，这块土地发生了变化。排水通畅的田地里长满了茁壮的庄稼，看不到一株杂草。牧场如今被修葺一新的防牲畜篱笆和新的大门围住，饲养着大量健壮的牛羊。有一天，农夫正靠在大门上欣赏这一切，牧师过来了。牧师说："你和上帝在这里造了一座令人赞叹的农场，约翰。""也许是吧，"约翰回答说，"但你应该见一见上帝独自拥有这里的时候是怎样一副惨状。"

　　早在旧石器时代，狩猎采集者们就知道，土地本身就会为人类生产食物，但他们的人口密度很低。为了养活现在的人类，农民利用土地，并在土地上投入自己的劳动力，动物，建筑，工具和机械，从化肥、农药到电脑和手机等购自其他行业的各种产品，以及兽医、顾问和科学家的专业服务。本章讨论的就是这些投入。但首先是土地。

土 地

没有土地也能生产粮食。农作物可以在营养液中生长,这种技术被称为水培。越来越多的沙拉作物在温室中种植,其中大量的产品来自面积相对较小的土地。猪和家禽等规模化禽畜也被安置在相对较小的土地上,当然生产它们的饲料需要更大的土地面积。但对大多数农民来说,土地,即英国经济学家大卫·李嘉图所说的"土壤原始而不可摧毁的力量",是基本的必需品。

世界土地面积刚刚超过130亿公顷,其中37.6%,也就是将近50亿公顷,是农业用地。其余的土地大约平均分摊给了林地和其他用途,后者包括城市土地、无法开垦的陡峭山地和冰雪覆盖的土地。在农业用地中,28.3%为耕地,3.1%为咖啡或苹果树等永久作物,其余的是种类和质量各不相同的草地与牧场。

世界上的大部分粮食来自农业用地,但农业用地的分布远不均匀。美国是世界上最大的农业出口国之一,那里只有44%的土地用于农业,而孟加拉国的农业用地超过70%,日本却只有12.6%。人均耕地数量也有类似的差异,澳大利亚人均有2公顷,日本却只有0.03公顷。2010年的世界平均水平是每人0.22公顷,是1960年的一半。这意味着在过去五十年中,每公顷的粮食产量肯定有所增长,事实上也的确如此,每年增长2%～4%。同时,耕地的数量也增加了,每年增加大约1%,但尚未耕种的潜在耕地分布很不均匀。据联合国粮食及农业组织(FAO)统计,南亚、西亚和北非没有"多余的"土地,可利用土地的90%集中在七个国家:刚果、安哥拉、苏丹、巴西、阿根廷、哥伦比亚和玻利维亚。由于其进一步可利用性将对环境和社会产生影响,我们难免怀疑那里大概根本没有多余的土地。

从第一章应该可以清楚地看到,并不是所有的土地都有同样的产量。土壤、海拔、朝向(坡面是朝北还是朝南——朝南的坡面在北半球获得更多的太阳辐射,反之亦然)和气候,特别是温度和降雨量方面的差异,都会影响生产率。

例如,从全球来看,坦桑尼亚的农业用地充足,人均0.77公顷,但其中大部分土地是干燥的稀树草原,只适合粗放式的放牧。相比之下,正如前文所述,日本的农业用地很少,但其中大部分土地都得到了集约化耕种,生产水稻和蔬菜。坦桑尼亚的人均土地面积是日本的20多倍,而日本每公顷土地的附加值却是坦桑尼亚的67倍。于是人们便想出了根据土地的灵活性对土地进行分级的方法:最好的土地几乎可以生产一切,而随着土地质量的下降,对土地生产的限制也越来越大。

水往往是限制因素。在某些情况下,土地过于湿润,或在一年中的错误时间成为需要排水的湿地。更常见的情况是土地过干,需要灌溉。有些作物(例如某些类型的水稻)完全依赖灌溉,世界上干旱地区的作物生产也可能要仰仗灌溉。

加利福尼亚州的中央谷地和落基山脉以东的高原干旱地区都有著名的例子,后者部分地区的下层是奥加拉拉含水层[①],该含水层从南达科他州延伸到得克萨斯州西北部。它的使用既说明了灌溉的潜力,也解释了其中的隐患。1950—1980年期间,由于大流量泵和中心枢纽灌溉系统的发展,含水层灌溉的土地面积从大约80万公顷增加到500多万公顷,由此产生的绿色作物

---

① 位于美国大平原之下被沙子、淤泥、黏土和砾石包围的浅潜含水层。该含水层是世界上最大的含水层之一,面积约为45万平方公里。

圈在该地区的卫星照片上仍清晰可见（参见图4）。到1970年代末，这个曾经饱受干旱之苦的尘暴地区开始出产玉米、棉花、小麦和高粱，但当时奥加拉拉含水层的排空速度显然远远超过了它的填充速度，正如加利福尼亚州的部分地区发生的那样，它的地下水位正在下降。

中国华北平原和印度的印度河-恒河平原也有不少不可持续开采的例子。这催生了类似于碳足迹的"水足迹"概念：为了提供我们从中获得的营养物质，不同的作物和动物产品需要的水量大不相同。全球作物生产中使用的大部分水（根据一项计算，高达78%）是雨水，这在那些计算中被称作"绿水"。"蓝水"是地表水和地下水，它们要么蒸发，要么流入大海，要么用于农业。它提供了大约12%的农作物生产用水。第三类是"灰水"，这是同化一切污染物所需的水量，占用水量的其余10%。

农场动物也要用水，显然是为了饮用，但还有大量的水被用于清洁它们的房舍和提供它们所吃的食物。总的来说，农业部门占全球蓝水消耗量的85%左右。表9显示了各种能量、蛋白质和脂肪来源在水足迹方面的巨大差异。

表9 每单位养分的水足迹

能量（升/千卡）	蛋白质（升/克）	脂肪（升/克）
牛肉 10.19	坚果 139	淀粉类块根 226
绵羊/山羊肉 4.25	牛肉 112	豆类 180
鸡肉 3.00	猪肉 57	牛肉 153
糖料作物 0.69	蛋 29	奶 33

续 表

能量(升/千卡)	蛋白质(升/克)	脂肪(升/克)
谷物 0.51	豆类 19	猪肉 23
淀粉类块根 0.47	油料作物 16	油料作物 11

数据来源：M. M. 麦肯宁和A. Y. 胡克斯特拉，《农场动物产品水足迹的全球评估》，《生态系统》第15期，2012年，第401—415页。

劳动力

在人类历史的大部分时间里，大多数人都生活在城镇之外。但自2008年以来，世界城市人口一直大于农村人口，据预测，城市家庭将占未来人口增长的大部分。这显然会对农业产生影响，因为大多数农业工人生活在农村，但必须记住，严格区分农业工人和非农业工人是第一世界的观念，并不被发展中国家所认可。在发展中国家，许多城市居民——例如布基纳法索首都瓦加杜古40%以上的人口——仍在从事农作物和牲畜的生产。

从整个世界来看，超过四分之一的人口（即从事经济活动的人口及其不工作的受扶养人）受雇于农业，但在这一总数中，各国之间存在很大的差异。2010年，整个非洲的数字是49.1%，印度是48.4%，拉丁美洲是15.8%，而美国、加拿大和英国的数字却只有人口的1.8%左右。然而，引用这些国际统计数字，却提出了一个看似简单但并不容易回答的问题：谁应该被纳入农业工人的范畴？什么才算农场？

在本书中，我们将从事农业经营的人称为"农民"，但在这一点上，我们应该认识到，使用这个词涵盖各种人群所从事

图6 加纳上西区索莫阿村的农民在市场上出售他们包括谷子（图中右侧）在内的农产品

的活动。当然，大多数国家的大多数农场都由个人或群体所拥有，并且主要使用家庭劳动力（图6）。换言之，它们是家庭农场。

尽管如此，还有一些人虽然不是农户家庭的成员，却也参与农业。最明显的是，有一些人在农场工作，但没有入股。他们被称为农场工人或农业工人，提供有偿劳动。相反，有些人拥有农田，但自己并不耕种。他们是地主，把自己的土地租给佃农以获取租金。有时，政府、宗教机构或企业会充当地主。另外，企业（有时被称为"综合农企"）或合作社可能会雇用管理人员和工人来耕种土地，而不是将土地租给家庭农民。很可能会有这样的农场：一个家庭拥有其中一些土地，又租了一些土地，并有权与其他农民共同使用其他的土地,例如放牧。

随着时间的推移,这些不同的角色往往获得了带有隐含地位的标签。因此,小型家庭农场主往往被称为农民(peasants)。在一些国家,这个词有以独立和自给自足为荣的含义;在另一些国家,它可能意味着地位低下和缺乏教养。不事耕种的土地所有者有时被称为地主,这个词本身似乎意味着地位高。相反,无地的农业劳动者心照不宣地被赋予了较低地位,他们包括那些教育程度低、收入微薄的流动工人,而在一些发达国家,则包括受过严格训练、高薪的农科毕业生。

我们或许能够想到,如果有许多种农场主,那么对农场是什么也有许多定义。乍看之下似乎很明显的是,农场就是一片土地以及相关的建筑和机械,人们在上面种植作物,饲养畜禽。但究竟是哪些人,有多少土地,种植和饲养的又是怎样的作物或畜禽呢?

联合国粮食及农业组织每十年进行一次世界农业普查,各国在答复普查时,对农场最小规模的看法各不相同。印度没有最低限度,而孟加拉国只包括超过0.2公顷的农场,中国则包括大于0.07公顷的农场。德国的最小农场面积为2公顷。在俄罗斯,98%的农场为私人所有,但其土地只占农田面积的2%;其他98%的农田为机构、承包企业家和非营利性公民协会所有。

因此,世界上农场数量的任何数字都不可能准确,因为它取决于定义,但最近对其进行估算的结论是,目前世界上有逾5.7亿个农场,而且由于低收入和中等收入国家的农场数量越来越多,这一数字还在增加。超过一半的农场在中国和印度,只有4%在高收入国家。其中大部分(逾5亿个)是家庭农场,即家庭拥有这些农场,并在其上从事大部分工作,而大多数农场规模较小,世界上94%的农场土地面积不到5公顷。

然而,通过世界平均水平来考察农场规模,却掩盖了很多地域差异。在东亚和南亚以及撒哈拉以南非洲,大多数农场的面积都在5公顷以下,它们至少占农业面积的一半。而在拉丁美洲,小农场较少,大部分土地都属于一些较大的农场。在高收入国家,虽然仍有许多小农场,但大部分土地由大农场耕种。例如在美国,按价值计算,80%的产出是由前10%的大农场生产的。表10列举了各国的一些例子。

表10说明了在讨论农场规模时遇到的一个难题:如果你向农民询问他们的农场规模,他们几乎总是以公顷、英亩或任何当地衡量土地面积的标准来回答,又或许以他们拥有的牲畜数量来回答,但这个问题问的实际上往往是农场业务的规模,以及它是否能够支持农场家庭。在非常肥沃的土壤上集约化种植的小农场,其产出可能远远高于只适合放牧几头牛的土地贫瘠的农场,尽管后者就土地面积而言是个大农场。

表10 20世纪末农场和土地的分布情况

	埃塞俄比亚	厄瓜多尔	英国	美国
农场数量(千)	10 758	843	233	2 129
农业总面积(千公顷)	11 047	12 356	16 528	379 712
农场规模(公顷)趋势	下降 1.4 → 1.0	下降 15.3 → 14.7	上升 55 → 71	上升 158 → 178
5公顷以下的农场所占百分比	99	64	23	11

续 表

	埃塞俄比亚	厄瓜多尔	英国	美国
5公顷以下农场的土地所占百分比	93	6	1	0.14
5～100公顷的农场所占百分比	1	35	60	38
5～100公顷农场的土地所占百分比	7	65	30	12
100公顷以上的农场所占百分比	0	1	17	27
100公顷以上农场的土地所占百分比	0	29	69	88
每个农场雇用长期工人的平均数量	0	0.3	0.6	1.4

数据来源:《联合国粮食与农业组织》,2014年,S. K. 劳德尔,J. 斯考特,S. 辛格,《我们对全球农场和家庭农场的数量与分布情况了解多少?》,《2014年粮食和农业状况》背景文件,农业食品经济司工作文件14-02号,罗马:粮农组织<http://www.fao.org/docrep/019/i3729e/i3729e.pdf>。经许可转载。

然而,国际统计组织只收集面积方面的农场规模数据,因此我们不得不使用那些数据。而厄瓜多尔29%的农业用地掌握在1%的100公顷以上的农场手中,这一事实明显告诉我们,厄瓜多尔农业产出的很大一部分可能来自那些农场。同样,英国绝大多数农场的土地面积不到100公顷,但与美国一样,大部分产出均来自那些拥有100公顷以上土地的农场。

这往往具有重要的政治影响。旨在帮助大农场的农业政策可能会使小农场主举步维艰。不过,无论是在政府选举中还是在农民利益团体的官员选举中,每个小农户都有投票权,因此,他们关切的东西在政治上可能比大农户关切的更加突出,尽管大农户可能有更多的时间和金钱用于政治活动。另一方面,在发达国家,即使是最大的农场也不如食品加工商和零售商的市场力量大,因此他们不得不和小农户一样接受报价。

表10说明的另一个显而易见的问题是,与大农场相比,小农场更可能依靠家庭劳动力,更不可能雇用劳动力。如该表所示,虽然英国和美国以及许多其他发达国家的农场规模日益扩大,但在发展中国家,随着人口的增加,农场规模往往在缩小。这意味着家庭的土地可能太少,无法有效地利用现有劳动力。如果当地的非农场农村经济有就业机会,这可能是一个有利的特点,但如果没有,则可能导致为了找工作而永久性或临时性地移居城市地区。

从国际统计的角度,甚或从国家农业决策者的角度来看,土地和以此为业的劳动力之间的关系已经够复杂了,但值得注意的是,我们所谈论的家庭不仅仅是耕种土地,他们通常在**自己的**土地上生活和劳作。他们可能已经在这块地盘上劳作多年了,在他们之前,他们的父母和祖父母可能也在这块地盘上劳作。因此,他们与土地的联系可能超出了纯经济的范畴。

欧洲农场主在被问及他们的商业目标时,可能会谈论利润最大化或长期增长,但其中也有许多人会说,他们的主要目的是将能够存续的农场生意传给下一代,就像他们接手的时候一样。据报道,当坦桑尼亚政府希望将马赛人的土地转为狩猎保护区时,一位马赛牧民代表也表达了同样的观点。他们得到了经济补偿,但"它不能和土地相比。土地是可以继承的。他们的母亲和祖母都埋在那片土地上。没有什么可以和这个相比"。随着家庭农场日益壮大，高收入国家更多的土地被农业公司耕种，这种情况是否会继续下去，还有待观察。

农场主和农场工人得到了许多其他专家越来越多的帮助。最初，农场主们也许利用农场的资源来满足他们的一切需要，铁匠协助他们制作犁铧，磨坊主协助他们加工谷物。现代农场主可能由专业承包商操作大型机器来犁地、播种、收割和除草。自己的机械发生故障时，他们可以从当地的机械经销商那里请技师来修理。他们的牛可以由巡回专家到农场进行人工授精，他们的育肥牛、绵羊和猪将由牲畜运输公司的卡车司机送往屠宰场。农艺师会告诉他们在作物上使用哪种杀虫剂，咨询师、顾问和银行经理会讨论他们的商业计划。在大学和研究机构工作的农业科学家或商业公司会为他们开发新技术。他们在农学院和大学接受培训时会了解这些，而之后由农业记者和广播员或农用化学品和农业机械行业公司的销售人员给他们介绍这些新技术。

他们的农作物和畜禽在离开农场后，会转入食品加工业和零售业。因此，虽然发达国家直接从事农业的人数在劳动力中可能只占极小的比例，但更多的人参与到了整个食品链中。

其他投入

人们最初开始栽培植物来吃，而不是简单地采集他们发现的野生植物时，他们需要某种挖掘工具来制备苗床，需要镰刀来收割谷物，还要篮子来收集它们。但大部分的工作及其背后的能量都来自人类的双手和肌肉。随着时间的推移，直接由人手完成的工作比例降低了，而农民使用的其他东西的重要性却增加了。

从役畜、种禽群和种畜群，到果树、建筑物、栅栏、篱笆、沟渠、排水道、工具、机械、化肥和农药，几乎数不胜数。它们都是农民在生产农作物、动物和动物产品的过程中使用的东西，其中大部分曾经都是在农场生产的。梳理苗床的牛拉耙子可能只是从树篱上剪下来的灌木那样的简易工具。奶牛群里的母牛和猪群里的母猪都是在农场繁殖的。用来做栅栏或建筑框架的木材可以从附近的林地或农场的树篱上砍来。树篱本身就是由农场员工种植的幼苗开始长成的，他们还在树篱旁边挖了沟。唯一的主要肥料是农场动物产生的粪肥，不过某些地方的农民可能幸运地生活在海藻等其他营养源附近。杂草是通过锄草来控制的。在一些发展中国家，这种方式还在继续。但在发达国家，现代设备已经基本取代了这种做法。也许最具代表性的工具是拖拉机，它在发达国家几乎完全取代了牛和马等役畜。

不同类型的农场需要不同的投入。在粗放式的牧场中，土地和劳动力占总成本的比例相对较大，因为农场工人的大部分时间都在照顾动物，并且土地很少施肥；而在高度机械化的集约化耕地农场中，机械、农药和化肥将占更多的支出。表11显示了不同国家在一些比较容易量化的投入方面的巨大差异。表中显示，富国往往比穷国使用了更多的采购投入，这是意料之中的，还显示出投入的使用也随土地耕作的强度而变化。

机械、农药和化肥用量（以及建筑等其他不易量化的投入）的巨大差异意味着农业的碳足迹因国家而异，也随着作物、禽畜产品或食品消费者的差异而不同。

温室气体主要有三种：二氧化碳（CO_2）、甲烷（CH_4，其全球暖化潜势为CO_2的23倍），以及一氧化二氮（N_2O，其全球暖化

表11　2000—2010年各国农业的采购投入

	每千公顷可耕地的拖拉机数量	每千名农业工作者所拥有的拖拉机数量	杀虫剂使用量（公斤/公顷）	化肥使用量（氮磷钾总量,公斤/公顷）
马里	0.2	0.1	0	7.4
坦桑尼亚	2.1	0.8	—	7.5
厄瓜多尔	4.0	4.6	3.7	88.2
巴西	12.9	28.4	—	111.7
印度	13.2	3.7	0.2	156.6
美国	27.0	707.1	2.2	107.7
法国	64.1	603.6	2.9	140.2
英国	75.6	976.4	3.0	238.2
日本	472.0	420.6	13.1	271.1

联合国粮食及农业组织，《2013年粮农组织统计年鉴：世界粮食与农业》，罗马：粮农组织，2013年。经许可转载。

潜势为CO_2的296倍）。二氧化碳是在为农机提供动力的内燃机制造化肥（特别是氮肥）的过程中，以及在林地垦荒或耕地时所产生的。反刍动物会排放甲烷，这是其消化过程的副产品之一，化肥和动物粪肥的分解会产生一氧化二氮。因此，使用大量化肥、高度机械化的农业会比欠发达农业产生更多的温室气体，但即使是低强度的放牛也会产生甲烷，对全球变暖产生一定的影响。

温室气体中农业所占份额的估计差异很大，从占总量的17%到32%不等。粮食及农业组织的一项研究发现，在全球人为排放中,仅世界上的畜禽就占二氧化碳的9%,甲烷的35%～40%,一氧化二氮的65%。如表12所示,不同食物的碳足迹也有很大差异,动物产品和进口商品的碳足迹高于植物产品。

表12　英国的温室气体排放量(每公斤产品的CO_2当量)

牛肉	68.8	咖啡	10.1
羊肉	64.2	水稻	3.9
动物脂肪	40.1	葵花油	3.3
猪肉	7.9	小麦	1.0
禽肉	5.4	菜豆	0.8
蛋	4.9	马铃薯	0.4
奶	1.8	糖	0.1

数据来源:P. 斯卡保罗,P. N. 阿普比,A. 米兹德拉克,A. D. M. 布里格斯,R. C. 特拉维斯,K. E. 布拉德伯里,T. J. 基,《英国食肉者、食鱼者、素食者和严格素食者的膳食温室气体排放》,《气候变化》第125期,2014年,第179—192页。

注:挪威的数字要低得多,见D. 布兰德福德,I. 加斯兰,E. 瓦尔达尔,《减少农业温室气体排放的"扩大化"与"强化":挪威的启示》,《欧洲选择》第12期第3卷,2013年,第4—8页。

因此,每个消费者的碳足迹都随其饮食而变化。英国的一项研究发现,那些每天吃100克以上肉食的人(大约两根契普拉塔小香肠加两片烤牛肉)每年相当于排放了2 624千克的二氧化碳,而素食主义者的这一数字仅为1 391千克。(作为比较,一辆十年车龄的家庭小汽车每年行驶6 000英里的碳足迹为2 440千克。)农民和食品消费者未来是否会面临减少这些碳排放的压力?

切记,农场主不只使用**物件**来帮助他们耕种土地和照管动物，传统、经验和知识也很重要。了解农作物或动物的生长状况，或者及时发现问题并采取措施，是农场主日常工作的重要组成部分。许多农场主在成长过程中或在农业培训中了解了照管好畜禽或"正确对待土地"的重要性。经济学家会说，这是行业的社会和知识资本的一部分。与机械和建筑等其他类型的资本一样，它可以通过投资来增加，在这种情况下指的是投资于教育和培训或研究和开发。现在，许多国家至少有一所农学院或大学的农业系提供全日制课程和其他培训机会，有的国家会有一个教育网络。其中许多还开展研究，或与其他研究机构有联系。

大约五十年前，世界上大部分农业研究是在发达国家进行的，而现在的模式是，新兴工业化国家的研究支出所占比例越来越大。2008年，中国、印度和巴西的农业研发支出合计为76亿美元，而美国、日本和法国的总支出为92亿美元。现在说这将对这些国家以及发展中国家的农业产生怎样的影响可能还为时过早。北美和西欧农业产业的经验表明，有效的知识网络需要时间来开发和传播新技术。

农业系统

如果我们现在将第三章中关于农业产出的信息与本章所整理的投入数据结合起来，应该就能够看到那些差异的组合如何产生不同的农业系统了。

这正是一些农业学术专家多年来为之着迷的事情，他们已经做出了相当复杂的各种分析，再加上从中学到的经验教训，可用于农场管理和农业政策的制定。不过，其基本思想相当简单，我们无意间已在应用了。根据农场在连续变量的几个标准上所处的位置：集约/粗放、简单/先进技术、大/小规模、混合/专业化、热带/温带、靠近/远离市场等等，每个农场各不相同。第五章和第六章将探讨这些差异如何影响目前的农业，以及农场主如何处理他们今后必将面对的问题。

第五章

现代和传统农业

人们对现代农业和传统农业之间区别的看法可能取决于他们生活的地方和时间。生活在水稻产区的年长者如果还记得1950年代的情况，可能会认为传统农业就是在动物的协助下种植高秆作物，只使用有机粪肥。而以薯蓣为主食的人可能会想到手工劳动生产的、种植在小山丘或山脊上的作物，两茬作物之间有很长的休耕期。

伊利诺伊州的麦克莱恩县位于美国玉米带中心的芝加哥西南约一百英里处，在1950年代，这里的农业正在从一种传统向另一种传统转变。那是拖拉机取代马匹的时代，也就意味着为马匹种植的燕麦和干草可以被更多的玉米和大豆所取代，大约一半的农场拥有玉米采摘机。但耕耘机和锄头仍然是控制杂草的唯一方式，农场也仍是家族企业。然而，西边几百英里处的传统农业意味着放牛，以及拥有更大的土地，而在南边几百英里处，传统农业却意味着种植棉花的小佃农。

在1950年代初的欧洲，传统农业意味着家庭农场的混合农业。例如,英格兰低地的一座典型农场可能有一小群奶牛,其中有短角牛、艾尔夏牛或娟姗牛,不需要当作后备种畜的小牛被饲养来生产牛肉,还有一群绵羊。在农场建筑的某个地方会发现几头猪,鸡和鹅会在农场上游荡。田地里种着谷类作物,通常还有几英亩的马铃薯,有时也种甜菜。也许有四分之一到一半的大片土地都是草地,为牛羊提供夏季的牧草,以及作为冬季饲料的干草作物。还会有几英亩的块根作物,如芜菁甘蓝、芜菁或饲料甜菜等,也是冬季的饲料。

机械化不会像在麦克莱恩县那样走得那么远,但通常会使用机器把牛奶挤进桶里,在牛舍中四处搬运。像小灰弗格森[①]这样的小型拖拉机越来越受欢迎,联合收割机也开始取代收割束禾机。但应该记住,这是1950年代的年轻人心目中的传统农业形象。当时的中年农民会认为马匹和手工挤奶是传统的,而拖拉机和挤奶机是现代化的。如果我们回到他们出生前一百年,回到19世纪初,会看到约翰·克莱尔[②]在他的诗《传统》中哀叹产生英格兰中部地区错落田野的圈地过程。他习惯于在开阔的田野里共同劳作,在那里

> 牛群来来去去,从傍晚到黎明,又到晚上,
> 以它们正当的权利走向野外草场。

并抱怨说

---

① 英国机械师和发明家哈里·弗格森(1884—1960)设计的农业拖拉机。
② 约翰·克莱尔(1793—1864),英格兰诗人。他是庄稼汉之子,以对英国乡村的赞美和对乡村遭到破坏的哀叹而闻名。

> 圈地来了，践踏着劳动权利的坟墓
> 穷人被迫为奴。

因此，即使被视为传统的东西也会随着时间而改变，人们还是经常将农业与传统相提并论。按照这种观点，农业被视作一种生活方式，在这种生活方式中，正确对待土地，生产健康的作物和畜禽，雇用当地人，把蓬勃发展、维护良好的农场传给下一代，比扩张、利润最大化和整合食品链更重要。那里并没有被异化的城市工人所特有的工作与家庭、劳动与休闲的分离。全家人都参与到农场中，全身心地投入工作，与那片土地和当地社区保持着密切的关系。他们的工作有着独立性和尊严。农民有一种认同感，这种认同感来自他们共同关注的问题、对农业报刊的阅读、农民组织的成员资格、共同的休闲追求，甚至来自他们的衣着风格以及对朋友和婚姻伴侣的选择。

显然，这是一个理想化的形象，但这是一幅由来已久的、有力量的画面。它是从城镇或宫廷角度来描写的源远流长的文学形式"田园诗"的一个版本，其中的乡村是另一番景象，诚实、和平和淳朴的乡村美德与机智、精明的宫廷德行形成了对比。

也许它更准确的叫法是农事诗，因为它把乡村生活理想化为一种令人满足的工作而不是无所事事。在美国，这种"杰斐逊式农论"有着悠久的历史，其基本观点在于农民是有着理想社会价值观的最可贵的公民。它源自人们对年轻时乡村的记忆，又被书籍、电影（想想电影《小猪宝贝》中农舍的画面）、照片和绘画（康斯太勃尔的《干草车》也许是个经典的例子）中的形象所强化。它也忽略了形象的消极方面——对陌生人的怀疑，以及自杀和工伤事故的高发率。与现代化、机械化、全球化的大型综合农企相反，它是可持续的，造就了野生动物的栖息地和美丽的风光，并关心动物的福利。这种对比的准确性如何，我们将在本章的以下部分进行研究。

可持续发展

发达国家的传统农业在多大程度上是可持续的，这一点一直存在争议。正如我们在前文中看到的，传统是什么不仅取决于我们讨论的地点，也取决于时间。20世纪中叶在北美和欧洲进行的那种农耕，现在的几代人可能认为代表传统，而当时尽管没有达到如今的程度，却也已经在使用以化石燃料为动力的拖拉机和化肥了。拖拉机从1930年代开始产生重大的影响，一些欧洲农民从19世纪中期就开始使用人工肥料，特别是磷酸盐。

19世纪末，欧洲动物生产的扩张严重依赖从美国和加拿大进口的廉价饲料谷物（美国和加拿大也提供面包谷物），所以如果我们想回到欧洲完全靠可再生资源为生的时候，就必须回到19世纪中叶，甚至更早。当时欧洲的人口约为2.6亿，而如今的人口接近之前的三倍。

1987年，世界环境与发展委员会为联合国大会编写了一份报告，通常以该委员会主席格罗·哈莱姆·布伦特兰之名称之为《布伦特兰报告》。报告将可持续发展定义为：在满足当代人需要的同时，不损害后代人满足自身需要的能力。因此，如果我们这一代人的活动没有为子孙后代留下足够的能源、水或肥料储备，造成全球变暖，环境污染，或破坏了野生动物的栖息地和景观，这些活动将被确定为不可持续。

这显然对农业有影响。不同的作物和动物生产企业具有不同的污染效应,正如我们在第四章讨论碳足迹时看到的那样,不同的作物和动物生产企业产生的温室气体排放量差异也很大,这意味着从气候变化的角度来看,有些作物和动物生产企业比其他的更具有可持续性。

但是,还有另外一个可持续发展的视角:农业所使用的有限资源是否有足够的储备来满足后代农民的需求?这个问题适用于所有种类的投入,从农民一直使用的土壤和水,到作为发达国家现代农业不可分割的一部分的拖拉机燃料、电力、化肥和杀虫剂,这些投入在发展中国家也越来越多,且颇具争议。

对现代农业的批评称,美国生产一吨玉米作为动物饲料需要一桶(42美国加仑或159升)石油,平均一公顷农作物所使用的化肥和农药需要两桶石油,以至于农业的能源使用量占美国的7%。另一位作者从不同的角度指出,农业的能源消耗仅占全球的3%,并认为在稀缺时期,石油往往被保留用作重要的用途,其中之一就是粮食生产。

另一种重要且有限的资源是磷。正如我们在第一章中所看到的,它是农民使用的肥料的三种主要成分之一。在每年开采的1.45亿吨磷中,大部分都用于生产肥料,根据联合国粮食及农业组织的估计,以这种速度,按照目前的价格和现有技术,值得开采的储量将在约82年后耗尽。然而更应记住的是,使用的速度会有很大的差异,而且开采技术也会发生改变,因此,对于世界何时达到"磷峰值"存在着很大的分歧。同样,对于另一种主要的化肥成分——钾,美国地质调查局的计算结果是,按照目前的消耗速度,95亿吨的可利用储量足够287年使用。

第三种主要的无机肥成分氮的情况不同。液化空气后再蒸馏分离出氮气相对容易,但要制成硝酸铵或尿素等有用的肥料,氮气首先需要与氢气结合制成氨气。这一过程中常用的原料是天然气,而制造氮肥的用量约占全球天然气年消费量的5%。

杀虫剂的化学成分更加复杂,使用的总量也较小,不会造成类似于化肥的资源需求,但与化肥一样,杀虫剂也会产生污染,这限制了它们的使用范围。一项关于美国农业对人类健康、水、土壤、空气质量和野生动物造成损害的成本估计表明,2002年的数字范围在57亿～169亿美元之间,另外还有37亿美元用于政府在监管和减轻损害方面的支出。

因此,为了实现可持续发展,无论是在全球范围内还是在地方规模上,现代农业都需要有农业质量合理的足够土地,土地上的水既不能太多,也不能太少,并且要有可靠的能源、化肥和农药供应,在不产生重大环境问题的情况下使用。人们常说,传统的农耕方式能自己生产能源和肥料,因而是可持续发展的。能量来自人或役畜的肌肉,这两种能量都是由农场生产的食物提供的。肥沃的养分在动物(通常也有人)产生的农家肥中循环。病虫害通过选择抗病品种、机械除草、休耕、轮作等方式加以控制,或者作为问题予以接受。因此有人认为,这就是我们必须在其还存在的地方维持的农业,或者从长远来看,我们必须回归这种农业。相反的论点是,这种农业方式生产的食品太少,不足以养活当前世界的人口。根据一项估计,世界人口的五分之二都以靠氮肥生产的粮食为生。

然而,现代农业可以变得更可持续。其中一种方法就是采用有机耕作的方式,不使用杀虫剂和无机肥料。就谷物而言产量较低，有时只有常规农业的一半，但就奶牛而言，产量却并没有低多少。发达国家的大多数有机农户仍然使用拖拉机，所以并没有减少化石燃料的使用。这种形式的耕作通常只占土地的一小部分：例如在英国，2004年，有机耕作土地约占农业用地的4%。

但还有其他办法。自1990年以来，面对可利用的投入的不断变化，古巴做出了最令人吃惊的反应。1980年代，古巴的农业以大规模的糖园为主，产品销往苏联和其他东欧国家。古巴农业严重依赖进口化肥、农药、石油和谷物。随着1989—1992年间苏联政权的解体，以及美国同一时期实施的贸易禁运，古巴的主要市场消失了，它再也负担不起重要的进口产品。古巴的反应异常迅速。几年内，国营农场部门就被移交给工人经营的合作社，规模也小得多，越来越多的粮食作物和农场动物出现在城市里，农民放弃了缺燃油少零件的拖拉机，回归畜力，合成肥料与农药被堆肥和动物粪肥、抗病品种、轮作和间作所取代，并使用捕食性昆虫来控制虫害。

从某种意义上说，所有这一切都只是对市场调节作用的反应：投入价格上涨，农产品价格也上涨，因此农民有更大的动力生产粮食，但使用的投入全然不同。政府也发挥了作用，改变了土地持有模式，利用古巴完善的农业科学家网络，并向消费者提供粮食价格补贴。

杀虫剂刚出现时显然有使用过度的趋势。例如在1960年代，沙巴（马来西亚）在新种植的可可作物上使用有机氯杀虫剂，这同时影响了害虫及其天敌，结果是害虫增多，作物被毁。停止喷洒后用寄生蜂控制了部分害虫，其他的害虫则通过清除作为其自然宿主的树木来控制。

类似的故事不胜枚举，它们让农民和农业科学家认识到，害虫和病原体可以迅速进化出抗药性，对其生命史的详细研究往往可以提出控制方法，通过各种控制方法将低水平的虫害袭击控制在一定范围内，往往比试图完全消灭害虫更好。

在发达国家，化肥和农药价格的上涨以及农业政策的变化也鼓励农民寻找更具可持续性的耕作方法。他们不再撒颗粒状的肥料，因为其中的一些肥料难免会落入树篱和沟渠中，而是喷洒投放更准确的液肥。英国推行了硝酸盐脆弱区，要求农民只在农作物生长的时候施肥，这样可以减少地下水位的损失。有智能手机的人如今可以下载"农场废料"应用程序，对粪肥和粪浆的施用率进行直观评估，并计算其所提供的养分，因此节省了购买化肥的潜在费用。

现在，英国90%以上的谷物种植区都喷洒了各种除草剂、杀菌剂和杀虫剂，其中一半以上的区域每年喷洒四次以上，但施用的有效成分总量却从1990年的1550万千克下降到2010年的940万千克。这些变化反映了社会对一些人可能视为农业副产品的宜人景观和野生动物栖息地的重视，下面一节将更详细地探讨它们与农业的关系。

野生动物与景观

美国生物学家蕾切尔·卡森是最早提请人们注意现代农业对野生动物的潜在影响的作家之一。她设想，当杀虫剂消灭了鸟类赖以生存的昆虫和杂草时，鸟儿再也不会在那里歌唱。这就是她为什么为1962年出版的书取名"寂静的春天"。从那时起，许多其他作家都提请人们注意现代耕作方式对环境的影响。

在热带国家,人们主要关注的是雨林的消失及其对全球变暖的影响。根据联合国粮食及农业组织2011年的一份报告,1990年代,亚马孙、刚果盆地和东南亚的森林每年损失710万公顷(总共13亿公顷)(图7)。2000年代,每年下降的数字为540万公顷,其中100多万公顷用于增加大豆的种植面积。在南美洲,其余的下降大都是牛场扩张的结果。

在温带国家,农业同样影响了野生动物及其栖息地,也影响了景观。这两者显然是联系在一起的,因为农业景观就是野生动物的栖息地。但农民很有可能以一片片的草场和摇曳的玉米丛创造出宜人的景观,却不能为野生动物提供什么食物或庇护所。同样,在大多数人看来没有什么景观价值的一片无法穿透的杂草丛生的区域,却有可能为各种哺乳动物、鸟类和昆虫提供

图7 农田取代了亚马孙雨林

食物和栖息地。因此，下面的讨论将在温带农业对野生动物和景观的影响之间进行看似相当人为的区分。

正如罗德里克·纳什在其经典著作《荒野与美国思想》中指出的那样，野生动物和荒野的概念本身就有赖于农业的存在。如果没有因为人类的播种而在田地里生长的植物，"荒野"一词就毫无意义；而在能够区分驯化的动物与野生的动物之前，动物也只是动物而已。因此，农业从一开始就对自然界产生了影响，虽然现代农业显然对野生动物产生了重大影响，但应该记住，这并不是什么新鲜事。

例如，英国女王伊丽莎白一世的政府在1566年通过了《保护格雷恩法案》，该法案到1863年仍然有效，它规定了对消灭那些公认为有害生物的鸟类和哺乳动物的悬赏奖金：一只红腹灰雀、十二只椋鸟或三只老鼠的头获赏一便士，一只鼹鼠的头是半便士，如此等等。另一方面，农民的活动也为野生动物提供了栖息地，从树篱形成的人工林地边缘，到旧谷仓中猫头鹰和蝙蝠的巢穴。

然而在过去的五十多年里，由于温带国家现代农业技术的影响，野生动物栖息地的范围和种类、野生植物的种类和动物数量都有所减少。传统的晚割干草牧场上的草长得很高，庇护着山鹑、黍鹀和长脚秧鸡，而现代的青贮草场则割得更短也更早，其成分主要是牧草，很少有传统草场上盛开的花。现代机械在大片田地上工作的效率最高，所以绵延数英里的树篱遭到铲除，随之而来的是巢穴和花朵的消失。在1950年代和1960年代，有机氯杀虫剂对猛禽繁殖成功率的影响是众所周知的。它们在西欧和北美的数量因此显著下降，游隼在落基山脉以东绝迹了。

最近，1990年代才得到广泛使用的新烟碱类杀虫剂被认为是蜜蜂和云雀等食虫鸟类数量减少的原因。这些只是一些比较著名的例子。

从第一批农民为农作物开垦土地开始，农业就一直影响着景观。航空公司乘客在美国中西部各州上空飞行时看到的棋盘图案的农场和方形田地是19世纪政府决策的结果。西欧的小规模农耕景观则更是渐进发展，中世纪的开阔田地在不同的时期被改造成较小的封闭田地，近年来这些田地的规模又有所扩大，而与传统风格的农舍和谷仓相配套的往往是新的工厂式建筑。

此外必须认识到，农耕活动也会在短期内改变地貌。犁过的耕地随着谷类作物的发芽而变成绿色，然后随着收获的临近而变成金色的谷浪。直到20世纪中叶，由于罂粟生长在玉米中间，谷田也可能是红色的，但现在选择性除草剂的使用增多，实际上已经从景观中消除了这种颜色。作物在春季或初夏开花时，绿色的大片油菜（油菜籽）会变成鲜艳的黄色，然后随着植物结籽而消退成暗褐色。甚至田地里牛的毛色也会因为农民的决定而改变：六十年前在英国乳品区出现的棕色或棕白相间的短角牛和艾尔夏牛，现在几乎完全被黑白相间的弗里斯兰牛和荷斯坦牛取代了。

动物福利

动物福利问题与上一节讨论的环境可持续性问题一样，也更容易在粮食丰富的发达经济体而非粮食匮乏的国家提出。

在农场饲养动物意味着我们接受这样的哲学立场，即为了人类的利益而饲养其他动物无可非议。必须认识到,有些人采取的立场是,人类为了获取食物而杀死动物是错误的,而另一些人则更进一步,认为任何形式的动物剥削,如为了牛奶或羊毛而饲养动物,都是无法接受的。显然,饲养动物的农民(以及食用动物及其产品的消费者)并不持这种立场,但他们大概都会同意,虐待动物是错误的。

更困难的问题是确定何谓虐待。在传统的小规模农场,工作人员熟悉每一个动物个体,应该能够确保畜禽的福利;大量的动物封闭饲养更容易造成生理和心理上的匮乏与疾病问题。这是对现代养殖业潜在的动物福利问题最简单的认识,这种说法表明问题有不同的方面:动物在多大程度上不能表现出正常的行为,规模的影响,以及疾病问题。

任何人看到圈养的动物在安顿下来吃草之前会绕着放牧的场地奔跑,都会意识到它们不仅仅是生物机器。看到沙浴的母鸡得到了明显的快乐,说明它的福利可能不仅仅包括足够的食物和饮水。相反,在舒适的垫料上避寒并躲避风雨,或者在炎热的气候下避开正午烈日而饲养的动物,大概会比没有这些保护的动物过得更好。

传统的养殖方法仍可能涉及诸如去角、去势和烙印等明显影响动物福利的做法,哪怕只是短暂的影响,尽管诸如鸡去喙(即切掉鸡的喙尖,让它们不易相啄)和切除猪尾的做法(这两种做法都是为了解决畜禽的密集型安置的问题)现在已经少见了。此外,圈养也有程度之分。群养在大型围栏里的动物不同于产仔箱里的母猪或层架式鸡笼里的母鸡,提倡改善畜禽舍条件的人指出,产仔箱可以防止母猪趴在后代身上,而应激的动物产量会下降，且易患疾病，所以没有一个农场主会刻意制造这种应激。

  这就是第二个因素经营规模的重要性所在。管理一个拥有数百或数千只动物的大型畜禽群所需的专业水平与管理传统的小群畜禽大不相同，而且更难对每个动物个体给予单独的关注。农民早就知道，大量的动物养在一起，疾病也就更容易传播。因此就有了反映农业智慧的老话，如"一只羊最危险的敌人是另一只羊"。

  高产也会给动物带来额外的压力。奶牛的乳房大，产奶量才会高，这给它的附着韧带带来了额外的压力，它也更容易患乳腺炎（乳房感染）和蹄部疾病。在喂养高精料、低粗料日粮的育肥肉牛中，已经发现了肝脓肿。这些都是农场主面临的问题，但也是整个社会在反思我们要用什么样的养殖方式来生产食物时，需要面对的问题。

第六章

农业的未来

到目前为止，本书中关于政府参与农业的内容还很少，但各国政府以及国际组织——如联合国粮食及农业组织和世界贸易组织——都对农业感兴趣。许多非政府组织、利益团体和慈善机构也是如此。

在这一章中，我们将探讨农业政策制定者目前必须思考的一些最重要的问题：气候变化会有什么影响；转基因生物（GMOs）会有什么影响，使用起来是否安全；未来我们是否能够养活世界。但首先我们需要看看围绕农业政策展开的一些争论。

政府和农业

大多数工业化国家的政府都以某种方式支持其农业产业，但程度却有很大的差异。澳大利亚和新西兰的农民几乎什么补贴都没有，而挪威、冰岛、瑞士、日本和韩国的农民2011—2013年的收入中有一半到三分之二来自政府补贴。由于在大多数情况下，农业只构成工业化国家经济的一小部分，我们也许会问，他们为什么要支持农民,为什么往往会有政府部委专门负责农业,而没有比方说专门负责零售业的部门,因为零售业雇用了更多的人,对大多数发达经济体的贡献也更大。既然在资本主义经济中,市场应该能确保满足消费者的需求,那么这些社会的政府为什么要干预农业市场呢?

从逻辑上讲,答案一定是农业市场不能生产社会和政府想要的东西。这些市场的失效体现在几个不同的方面。首先,它们不善于应对价格波动。人们对农产品的需求是相当稳定的,因为食品消费者每天都要进食,而且数量大致相同,也许圣诞等节日除外。但这些产品的供应量却会因为天气和疾病的变化而发生很大的变化。正如我们在第三章中所看到的,其结果是价格的波动。高价令贫穷消费者生活艰难,因为他们必须将很大一部分收入用于购买食品。微利农场在平均价格水平下还可能长期生存,而低价可能意味着它们在经历一两年的低收入后就会被淘汰。因此,人们过去普遍认为,应对这种价格波动是值得的。问题在于控制价格波动的政策可能会变成提高农场价格和/或收入的政策,而这些政策的争议性更大。

这使我们想到了农业市场失效的其他方式。其中一个问题是,农民向其销售产品的公司通常比农民拥有更大的市场力量——通过自己的决策影响价格的力量,仅仅是因为它们规模更大。另一个问题是,市场通常不会为农民所提供的作为粮食生产副产品(以野生动物栖息地和宜人景观的形式)的生态系统服务买单。第三个问题是,工业化国家对农民生产的基本粮食商品的需求并没有快速增长(见第三章)。因此,如果农民采用新技术并增加产量,价格将趋于下降。此外,他们还可能面临其他国家农民的竞争,因为这些国家的农民能够以较低的成本生产并从中获利。

农场游说团体认为,工业化国家的政府应该解决这些问题。他们的论点是,一个国家应该能够养活自己,或至少保持合理的自给自足水平,这有几个原因。

一是国家安全。各国政府有责任确保在紧急情况下能够提供粮食,尽管可以说,稳定的贸易关系是一种同样有效的粮食安全形式,因为与进口商关心确保供应一样,粮食出口商也关心确保市场的准入。另一个论点涉及就业问题:农业部门的衰败不仅意味着农村工作岗位的丧失,还意味着供应农民的行业,如化肥和农业机械,以及加工农民产品的食品业的就业率也在下降。还有付钱购买进口粮食的问题。各国政府必须决定其国家资源最好是用于生产本国的粮食,还是用于生产可以出售的其他东西以换取粮食。

即使在权衡了所有这些考虑因素后,政府决定不干预农产品市场,农民仍可能辩称,他们所面临的收入下降和波动并非自己的过错。2000年,英国每个农民的收入只有1976年水平的四分之一左右,不过到2010年已经恢复到那个水平的一半多一点。既然在这种情况下,社会上的其他群体(比如养老金领取者和失业者)实际上都得到了国家的支持,农民难道不应该获得支持吗?另一方面,还有一些诸如小店主和钢铁工人等不同的社会群体,他们在传统或衰退的行业中工作,却很少或根本没有得到国家的援助,而且他们与业主经营的农场主不同,没有像农场这样的大型资本资产可以依靠。

这些观点表明,农场收入保障问题涉及价值判断,社会通常通过政治机制来解决这些问题。因此,政治利益团体,如英国的全国农民联盟,或美国的美国农场局联合会、国家农庄和全国农民联盟,或是欧盟的农业专业组织委员会,其功能就是为农民辩护。诸如代表食品业的团体或环保主义者等其他团体可能会有不同的论点。

工业化国家农业政策的历史表明,政府既要对有效的游说做出反应,也要对不断变化的经济和政治环境做出反应。在美国,对农业市场的干预可以追溯到1930年代的《农业调整法案》,该法案不仅允许联邦当局从农民手中购买粮食,而且还可以向他们支付不在部分土地上种粮食的费用。这就是约瑟夫·海勒在他的小说《第二十二条军规》中嘲笑的政策。小说中,梅杰少校的父亲因为不种苜蓿而得到报酬,所以他没有种更多的苜蓿,他用政府付给他的钱买了更多的土地,这样就能不种更多的苜蓿。

美国的农业政策关注的仍然是如何平抑市场波动。2014年《农业法》有一项规定,即市场价格低迷时,可以从公共财政中向农民支付各种款项;对于奶农来说,当牛奶价格与饲料成本之间的差距低于一定的触发水平时,也可以从公共财政中获得各种款项,此时,农业部长会在国内粮食计划中鼓动购买乳制品。

19世纪末20世纪初,英国的历届政府都依靠进口农产品,特别是从美洲和大洋洲进口的产品,直到1930年代末。随后,当时占英国粮食供应近三分之二的海上贸易受到来自水面军舰和潜艇的威胁。对此,政府启动了一系列价格补贴,以促进国内农业产量的增加。第二次世界大战后,由于英国经济已无力支付以前的进口水平,这些补贴得以维持,并成功地将国内的自给自足水平提高到1973年国内消费的三分之二左右。

英国当时加入了后来的欧盟,其共同农业政策(CAP)运作的方式与欧盟不同。它利用进口关税将国内价格维持在高于世界水平的价位,而剩余的农产品最初被官方商店买进,随后在出口补贴的帮助下销往世界市场。这一政策非常成功地实现了最初的设计目标,即增加欧洲农业的产出。但到了1980年代,由于其成本、对环境的影响,以及对世界农业市场的干扰,该政策受到了大量批评。因此,它发生了根本性的变化。

欧盟是世界上最大的农业进口方,并且遥遥领先,其进口的障碍已经减少,如今对农民收入的支持已经采用了直接付款的方式。这些款项并不鼓励他们生产比满足国内和出口市场的需求更多的产品,但它们取决于农场的业务规模,因此,大农户比小农户得到了更多的钱。在某些情况下,这导致非常富有的人从公共财政中获取资金;是否有理由认为这是在付款购买他们提供的环境服务,还要由读者来决定。

发展中国家和新兴经济体中,农业政策的种类与工业化国家的一样多,甚至更多,包括提高或稳定价格的措施、投入补贴、教育和技术培训投资、研究与开发、咨询工作、发展信贷机构和生产者合作社以及土地改革等。

问题的部分原因是,专家和政治家关于何为促进发展的最有效方式的看法随着时间的推移而发生了改变。五十多年前,人们认为,农业发展是整个经济发展的先决条件。后来看来,对教育和卫生的投资回报更快,所以一些国家的农业价格保持在低水平,投资集中在城市地区。最近,舆论又转向了促进农业发展。同时,关于促进农业发展的最佳方式,意见也发生了变化。

在1960年代,重点是研发以及对种子和化肥的补贴,后来则转变为试图确定最适当的价格水平,但最近又重新关注研究和开发,世界银行也主张对种子和化肥进行补贴。

气候变化

气候变化如今已经是人们广泛承认的现实,尽管2011年对美国艾奥瓦州1 276名农民的调查发现,32%的农民认为气候变化没有发生,或没有足够的证据来证明。政府间气候变化专门委员会(IPCC)预测,到2050年,全球平均气温将比工业化前的气候上升2～4 ℃。换句话说,1950—2010年期间,全球平均气温每10年上升约0.13 ℃,预计今后30年将每10年上升约0.2 ℃。

气温升高可能会增加水循环的活动水平,简单地说,这意味着更多的降雨。虽然季节性降雨的时间和模式似乎有可能发生变化,但降雨的地点和时间更难预测,特别是在季风地区。较高的温度也会影响极地冰盖和山区积雪的融化,预计到2100年海平面将上升两米。一个地区的气候变化也有可能对其他地区产生影响,最明显的例子是喜马拉雅山的积雪减少意味着印度河和恒河周围平原更加干旱了。除了这些渐进的变化之外,气候的多变和极端天气事件也可能增加,这意味着干旱更频繁、时间也更长,极端潮湿天气导致北半球洪水的风险增加,热带气旋减少,但时间更长、强度更大。

农业和气候变化有着双重的关系。气候对农业有影响,农业对气候也有影响。先说后者,我们(在第四章)已经讨论了农业的碳足迹,指出农业在世界温室气体排放中占了很大的比重,尽管对其确切的程度还存在争议。这些排放有很大一部分——有人估计高达75%——来自中低收入国家，因为这些国家更多地生产水稻，燃烧生物质。因此，农民面临着改变的压力，这既是对政策制定者的回应，也符合他们自身的利益。他们可能需要升级排水系统以应对增加的降雨量，或使其作物适应较短的生长季节或更频繁的干旱。为了减少温室气体排放，他们可以通过改变种植的作物或使用的耕作方法来增加土壤碳。他们还可以在厌氧消化器中处理动物和作物的废物来收集甲烷，而不是任其逸散到大气中，同时可以增加有机粪肥的使用以降低化肥消耗。

就对农业的影响而言，除了灾难性的风暴和洪水之外，全球变暖的预期影响中最引人注目的也许是海平面上升导致的农田丧失。在沿海地区，特别是孟加拉国、缅甸、印度、巴基斯坦和埃及的河流三角洲，据估计约有6.5亿人会受到影响。相反，喜马拉雅山的降雪量减少，预计会在旱季造成遥远的印度河、恒河和布拉马普特拉河流域缺水，而这些流域附近有5亿人的生活和农业用水依赖于此。在非洲，尼日尔河流域的水量在20世纪有所减少，2010年俄罗斯的干旱使小麦产量减少了40%。

气温升高和降雨模式的变化会限制生长季节的长度。例如在加纳北部，农民还记得雨季长到足以收获两次，而现在如果不下雨，第二茬作物就危险了。在整个世界范围内，约80%的农业面积是靠雨水灌溉的（非洲和拉丁美洲的比例更高），但其余20%的灌溉土地生产的粮食占世界粮食产量的40%以上，人们认为这些土地容易受到气候变化引起的缺水问题的影响。

温度波动，特别是谷物开花期的高温（32℃以上），也会影响作物产量。实验表明,超过这一温度,水稻产量会降低90%。玉米作物试验中,超过30℃的时间每多一天,在最佳降雨条件下最终产量就减少1%,在干旱时减少1.7%。

然而还有必要记住,较高的温度可能对温带地区有利,因为潜在的作物生产面积可能会增加,生长季节的长度会增加,降雨量也可能增加,而寒冷期则会减少。大气中二氧化碳含量较高也会对作物生长产生施肥的效果(见第一章)。因此,与撒哈拉以南非洲和亚洲相比,北美洲、拉丁美洲南部和欧洲的农业受到不利影响的可能性较小,一些作物可能会增产。

相反,气温升高也会影响害虫的生存和生长能力。事实证明,在二氧化碳浓度较高的地方,蚜虫和象鼻虫的幼虫生长速度较快,冬季气温较高会降低季节性死亡率,使害虫物种能够更早、更大范围地扩散。动物健康也可能受到影响。近来最著名的例子可能要数牛羊蓝舌病向北欧和西欧的蔓延了。蓝舌病病毒通过库蠓属的蠓虫叮咬传播,地中海地区的人们很久以前就认识到了这一点,残肢库蠓是其媒介昆虫。最近,其他种类的蠓(如不显库蠓和灰黑库蠓)似乎也感染了该病毒,而且由于冬季较温和,它们和病毒的生存能力都有所提高。因此便造成了这种疾病的传播,到2007年已经蔓延到了英格兰东南部。

鉴于可能产生的影响范围广泛,要预测气候变化的总体影响并不容易。对1980—2008年温度和降水影响的研究得出结论,温度的影响大于降水,与这一时期没有这种气候趋势的情况相比,玉米会减产3.8%,而小麦则会减产5.5%。水稻和大豆产量的增减大致持平,而温带地区的水稻产量则得益于较高的气温。各国之间可能存在很大的差异:俄罗斯的小麦减产了近

15%，而美国没有受到影响，因为那里没有明显的气候趋势。该研究的作者承认，他们的模型可能是悲观的，因为它忽略了二氧化碳的肥化影响。考虑到这些因素，他们估计，在这28年中，气候变化造成的商品平均价格总涨幅为6.4%。

然而，试图利用这样的研究预测未来困难重重。它们不仅忽视了农民适应不断变化的环境的可能性，而且还忽视了农民生活工作的更广阔世界的变化。因此，政府间气候变化专门委员会提出了气候脆弱性指数，该指数结合了17个不同的变量，不仅包括气候，还包括人口、社会经济和治理措施。这清楚地表明，非洲和亚洲的大部分地区是最脆弱的，而北美和欧洲大部分地区是最不脆弱的。

转基因生物

传统育种者的工作方式是试图将同一物种的不同变种或品种的理想性状结合起来。例如在20世纪初，剑桥大学的比芬教授将成熟的英国小麦品种"北欧佬大师"与俄罗斯品种"吉尔卡"杂交，后者对小麦植株叶面的真菌病——黄锈病——具有抗性，从而培育出新的抗锈品种"小乔斯"。同样，动物育种者将一个产奶量高的绵羊品种与另一个产羔率高的品种杂交，培育出能产大量奶水哺育双胞胎羔羊的种母羊。由此可见珍稀畜禽品种和外来植物品种的重要性。它们代表了具有不同特性的基因库，可供未来使用，以应对不可预测的需求变化。

这些传统方法的问题是耗费时间，因此培育新品种或变种的成本很高。苏格兰植物育种站（现为詹姆斯·赫顿研究所的一部分）于1937年开始了培育抗枯萎病真菌（见第一章）的马铃薯品种的工作。融合了抗枯萎病的墨西哥马铃薯基因的品种"彭特兰王牌"直到1951年才公布，据说这还是一个相对快速的育种方案实例。此外，这些常规方法也有局限性：如果一种性状（如抗枯萎病）不存在于某个品种基因中的某个地方，则无法发明出这种性状。从1920年代开始，农作物育种者就试图通过人工诱导突变来增加遗传变异，例如使用X射线或伽马射线。这是一种随机的技术，需要非常多的植物才能产生少数有用的个体，但它已被用于生产油菜（油菜籽）、水稻和大麦的新品种了。

转基因至少在理论上提供了一种将新性状融入作物的更快速也更精确的方法。这涉及几个阶段。首先，必须找到一个具有所需特性的生物体。接下来是在供体中分离出负责产生理想性状的基因，这是通过使用限制酶在所需位置切割其脱氧核糖核酸（DNA）来完成的。然后，在使用"基因枪"或细菌将该基因插入目标植物之前，对其进行克隆，即多次复制。使用基因枪是指将克隆基因附着在微小的金属颗粒上，然后在高压下插入目标植物细胞，使供体DNA整合到目标植物细胞核内的DNA中。这听起来很粗略，但已经成功应用了，特别是在小麦或玉米等单子叶作物上。

对于马铃薯、番茄或胡萝卜等双子叶植物，首选技术是使用一种天然的植物寄生虫——根癌农杆菌——将供体DNA输送到目标植物中。然后，对培育出来的植物进行测试，检查它们是否具有所需的特性，接下来扩繁它们并将其投放市场。

该技术的早期用途之一（至今仍然十分重要）是生产一种对草甘膦有抗性的大豆品种。草甘膦是一种除草剂，几乎可以杀死所有的绿色植物，以"农达"（RoundUp）的商品名出售。为孟山都公司（生产草甘膦的公司）工作的研究人员发现了一种对草甘膦有抗性的细菌，并将相关基因整合到大豆品种中，他们将该品种命名为"抗农达"（RoundUp Ready）。该品种在1990年代中期上市，农民能够在不影响作物的情况下，通过喷洒草甘膦来控制杂草（也增加了草甘膦的销售）。

另一个著名的例子是，土壤细菌苏云金芽孢杆菌（Bt）能够通过感染昆虫幼虫肠道的细胞膜而产生一种对其有毒的蛋白质。该菌的不同品系会产生不同种类的蛋白质，影响不同类型的昆虫。相关的基因同样可以确定。1990年代中期，它们被加入玉米和棉花的一些品种中，并提供了对玉米的主要害虫欧洲玉米螟，以及棉花的主要害虫棉铃虫和红铃虫的防护。随后的发展导致了性状的"叠加"，即现在有一些玉米品种同时具有Bt和抗草甘膦的基因。

基因改造也被应用于产生抗旱和抗病毒攻击的能力，并取得了不同程度的成功。水稻品种（所谓的"黄金稻"）被设计成维生素A含量更高，因此有助于防止失明。有一些大豆品种含有较高浓度的欧米伽-3脂肪酸，被认为有助于减少心血管疾病。基因改造已被更具实验性地用于繁育转基因农场动物。例如，来自菠菜和线虫的基因已被加进猪和奶牛的基因，提高了肉类和牛奶中欧米伽-3脂肪酸的水平，以减少人们患心血管疾病的风险。然而，到目前为止，转基因农场动物还停留在实验室内。

自转基因作物问世以来的二十多年里，其播种面积大幅增加。一位转基因倡导者称，全世界的转基因作物播种面积已经从1996年的170万公顷增加到2012年的1.7亿公顷，工业化国家和发展中国家平分秋色。其中大部分面积种植的是一开始就涉及的四种作物：大豆（现在世界产量的四分之三来自转基因品种）、棉花（近一半为转基因）、玉米和油菜（每一种都有超过四分之一的产量是转基因）。美国等一些国家的这一比例更高，大部分的大豆和油菜籽，以及相当份额的棉花和玉米，现在都来自转基因品种。

所涉及的作物范围也有所扩大，当前已经有了转基因版本的苜蓿、苹果、西葫芦、甜菜、甘蔗和小麦。现在夏威夷的番木瓜作物大多来自添加了抗番木瓜环斑病毒基因的品种。

然而，转基因作物仍然争议不断，大多数欧洲国家都禁止或谨慎控制其使用。反对转基因的人认为，将基因组测序和基因标记图谱等现代科学技术与传统的育种方法相结合，就提高产量而言，与基因工程一样有用。

本书没有足够的篇幅来详细讨论关于转基因的所有论点，因此只能列出它们。有人担心，有些人可能会对转基因品种产生过敏或有中毒反应。也许更重要的是担心改造过的基因可能会无意中传播开来，对非目标生物产生不可预知的影响。一项针对德国种植的转基因玉米的研究发现，很难防止被风吹走的转基因玉米花粉给非转基因玉米植株授粉。在南非和加纳进行的类似研究导致人们对传统玉米品种的维护感到担忧。

还必须记住，植物和动物对毒性的敏感程度天然不同，因此，杀死最敏感的个体对最不敏感的个体有利，这样一来，动植物物种往往会形成抗药性。路透社2011年9月报道，包括俄亥俄州的小蓬草和堪萨斯州东北部的糖果苋在内的21种具草甘膦抗性的杂草侵袭了美国1 100万英亩的土地。为了控制害虫物种抗性的发展，美国的农民被要求在部分作物区内种植非Bt品种,作为无抗性昆虫的避难所,但在世界其他地区,维持这一政策可能会很困难。孟山都网站将其在印度的一个Bt棉花品种的红铃虫抗性丧失归咎于当地农民没有种植非Bt庇护作物。

还有一些社会和道德方面的担忧。由于转基因品种通常每年都要从种子公司重新购买,因此保存种子的传统做法再也行不通了,这尤其会影响到贫穷的农民,而生产这些品种的跨国公司却从中受益。这些实际或潜在的问题将对目前转基因作物的推广产生何种影响还有待观察。

养活90亿人

2010年,联合国人口司的中期估计预测,2050年全世界将有93亿人。2013年,《纽约时报》的一篇文章计算,目前的农业为世界70亿居民每人生产2 700卡路里。如果不增加粮食产量,当前的产量仍然能为2050年的人口每人生产2 000多卡路里的热量,完全可以满足生存的需要。那么,还有什么可担心的吗?

如果你读到了这里,就会意识到事情并没有那么简单。目前,这些卡路里约有三分之一用于饲养动物,有些将在食品链中被浪费掉,现在约有5%将被转化为生物燃料。剩余的供给并不是平均分配的,因为工业化国家比发展中国家人均得到的更多,各地的富人都可以比穷人花更多的钱购买食物。不仅如此,未来人们的食物需求可能不同,生产粮食的资源和技术也会不同。本书的篇幅太短(更不用说这一章的篇幅已所剩无几),无法对2050年可能出现的世界粮食平衡做出深思熟虑的估计,但我们至少可以看看在这样做时需要思考的一些问题。

正如我们在第三章中看到的那样,影响未来粮食需求的两个主要因素是世界上的人口数量和他们需要花费的资金数额。93亿的中等人口估计数假定,2045—2050年,妇女人均将有2.17个孩子;2005—2010年,妇女人均有2.52个孩子。但如果妇女人均有2.64个孩子,即高估计值,就可能有106亿张嘴要养活了。因此,就未来的粮食需求而言,影响女性生育率的因素,如教育、收入、节育权、宗教和文化等至关重要。总的来说,随着社会更加富裕,人们往往倾向于少子。但也正如我们在第三章中所看到的那样,他们也会吃得更多,饮食更加多样化。我们已经看到,随着中国经济的增长,对猪肉的需求增加了,因此对饲料谷物的需求也增加了。预计其他新兴经济体随着消费者收入的增加,对动物蛋白和脂肪的需求也会增加。简而言之,我们应该期待粮食需求的增长超过人口增长,但增长多少则相当难以预测。

我们所知道的是,在过去,供给的增长远远满足了不断增长的需求,特别是不断增长的人口。据历史学家乔瓦尼·费代里科计算,1800—2000年期间,世界人口增加了六七倍,但同期世界农业生产至少增加了十倍。在19世纪,大部分的增长都来自北美洲和南美洲以及澳大利亚和新西兰的土地,这些土地以前没有为世界市场生产过粮食和纤维。20世纪的增加主要是每公顷土地和/或在土地上工作的人的产量增加的结果。这些解决方案中的任何一种都可以在21世纪再次奏效吗?

在阿根廷和巴西等国,我们仍在砍伐树木来种植玉米和大豆,但这受到了越来越大的政治压力,而且正如前文所述,随着全球持续变暖,我们很可能因海平面上升、干旱和荒漠化而失去土地。要详细预测这些趋势将如何相抵并不容易,但似乎显而易见的是,我们不能指望19世纪那种基于额外土地的产量增长。

这使得我们不得不在每公顷农田上生产更多的粮食。过去，面对需求相对于供应的增加，以及随之而来的价格上涨，农民善于通过额外的投入——从增加更多的除草人工到额外的化肥和杀虫剂——来增加产量。科学家和技术专家也善于培育更好的植物和畜禽品种，生产更好的机械等。1970—2009年，世界人口只增加了逾80%，但世界玉米产量却增加了三倍，小麦和水稻产量也增加了一倍多。这种情况能否持续到未来？

工业化国家增加产量的简单方法的范围可能很有限，但已经有了一批训练有素、经验老到的农业劳动力，他们习惯于采用综合病虫害管理等新技术，以寻求实现可持续集约化的理想。

增加每公顷产量的最大机会也在需求增长可能最大的地方：新兴和发展中经济体。在这些国家，多施一点肥料，或是使用稍微好一点的种子，或者多用一点牛或驴的畜力，就能发挥最大的作用。农民用手机既可以获得技术建议，也可以了解当地哪个市场的价格最好。即使是一些简单的改变，如附近有自来水供应，或一种不需要妇女每天花两三个小时寻找柴火的烹饪方式，也意味着有更多的时间为作物除草、种植第二种作物，或是照料动物。因此，使世界农作物和畜禽的平均产量达到生产力较高的农场已经达到的水平，是21世纪人类面临的挑战。

索 引

(条目后的数字为原书页码，见本书边码)

A

agribusiness 综合农企 75
Agricultural Adjustment Act (USA) 《农业调整法案》(美国) 103
agricultural advisers 农业顾问 79—80
agricultural education and training 农业教育和培训 80, 84
agricultural incomes 农业收入 102—104
agricultural markets 农业市场 60—64, 100—103

    agricultural prices 农业价格 61—64, 101—102

    demand for agricultural products 对农产品的需求 60—64, 101, 114

    market failure 市场失效 101—102

    supply of agricultural products 农产品的供应量 60, 63—64, 67, 114—116

agricultural policy 农业政策 67, 78, 100—105

    farm income support 农场收入保障 100—101, 103—104

    in developing countries 发展中国家的 105

agricultural pressure groups 农业利益团体 103
agricultural products 农产品 51—60; 参见 international trade in farm products

    temperate products 温带产品 65

    tropical products 热带产品 65—67

agricultural research and development 农业研究与开发 84
agricultural sustainability 农业可持续发展 89—94
agricultural tools and machinery 农业工具与机械 69, 82, 86—87, 92—93
agricultural workers 农业工人，见 labour
alfalfa 苜蓿 54
alpacas 羊驼 32
American Farm Bureau Federation 美国农场局联合会 103
amino acids 氨基酸 35
ancillary workers 辅助工人 79—80
animals, farm 农场动物 32—50; 参见各个物种；livestock

    animal behaviour 动物行为 49

    animal breeding 动物育种 39—44, 111

    animal health and disease 动物健康与疾病 45—47, 108

    animal husbandry 畜牧业 34, 48—50

    animal nutrition 动物营养 34—39

    animal production systems 动物生产体系 47—50

    animal products 动物产品 33—34, 58—59

    animal traction 畜力 33

    animal welfare 动物福利 45, 97—99

antibiotics 抗生素 46
apples 苹果 60
arable farming 耕作 2, 27—31
artificial insemination (AI) 人工授精 43
artificial lighting 人工照明 44

B

Babe (film)《小猪宝贝》(电影) 88
Bacillus thuringiensis (Bt) 苏云金芽孢杆菌 111, 113
bananas 香蕉 23, 66
barley 大麦 2, 27, 29, 30, 55
battery cages 层架式鸡笼 1, 99
beans 菜豆 12, 27, 29, 30, 56
Biffen, Professor Roland 罗兰·比芬教授 109
biodiesel 生物柴油 57
biological control of pests and diseases 病虫害的生物防治 26, 93
branding 烙印 98
brassica crops 芸薹属作物 27
Brundtland Report《布伦特兰报告》，见 World Commission on Environment and Development
buffalo 水牛 33, 36, 43, 58
butchery 屠宰业 53

C

cabbages 卷心菜 27
California (USA)（美国）加利福尼亚州 2, 71—72
camels 骆驼 33, 35, 36, 58
capital 资本 80—82
 social and intellectual capital 社会和知识资本 83—84
carbohydrates 碳水化合物 2
 carbohydrate metabolism in animals 动物的碳水化合物代谢 34—36
carbon cycle 碳循环 6—7
carbon footprint 碳足迹 81—83, 105—106
Carson, Rachel 蕾切尔·卡森 94
cassava 木薯 2, 16, 27, 30, 55, 59
castration 去势 98
cattle 牛 32—40, 58
 cattle breeds 牛的品种 39—40, 42—43
cellulose 纤维素 18, 34—36
cereals 谷物，见 barley; maize; millett; oats; rice; rye; sorghum; wheat
chickens 鸡，见 poultry
Clare, John 约翰·克莱尔 87
climate change 气候变化 90, 105—109
 Climate Vulnerability Index (IPCC) 气候脆弱性指数（政府间气候变化专门委员会）109
clover 白车轴草 12, 30
cocoa 可可 65—66
coffee 咖啡 65—66
Comité des Organisations Professionelles Agricoles (COPA) 农业专业组织委员会 103
Common Agricultural Policy (CAP) 共同农业政策 104
concentrated animal feeds 动物精饲料 38—39
Constable, John 约翰·康斯太勃尔 88
continuous cropping 连续耕作 11
corn 玉米 31；参见 maize
cotton 棉花 28, 30, 51, 54
cowpeas 豇豆 31, 57
cows, dairy 奶牛 38, 45, 47, 58, 87

crops 作物 16—31；参见各作物
    cereal crops 谷类作物 16—17
    crop growth 作物的生长 17—20, 23—27
    crop nutrients 作物的养分 12—14, 24
    crop pests 作物的害虫 2, 21—22, 26, 108, 111
    crop products 作物的产品 27—29
    crop rotation 轮作 26, 30
    crop varieties 作物的品种 19, 28—29, 109—112
    crop yield 作物产量 20, 26, 55—56
    diseases of crops 作物的病害 22—23, 26—27
    root crops 块根作物 16, 55
    tree crops 木本作物 28, 64, 70
    world area, production and average yield 全球种植面积、产出及平均产量 55—56
Cuban agriculture 古巴农业 92—93
cultivation 栽培 28—29

D

dairy industry 奶业 53
dehorning 去角 98
Diamond, Jared 贾雷德·戴蒙德 32
digestive anatomy 消化道解剖 36—37
donkeys 驴 33, 36
drainage 排水 9, 71
ducks 鸭，见 poultry
Dust Bowl (USA) 沙尘暴（美国）4—5, 11

E

ecosystem services 生态系统 52, 101
eggs 蛋 1—2, 40, 66
    free range 散养 2
embryo transfer 胚胎移植 43—44
energy consumption by agriculture 农业的能源消耗 90
ethanol 乙醇 57
European Union (EU) 欧盟 1, 66, 104

F

family farms 家庭农场 74, 76
farm animals 农场动物 32—50
farm capital 农场资本，见 capital
farm labour 农场劳动力，见 labour
farm lobby groups 农场游说团体，见 agricultural pressure groups
farm size 农场规模 75—78
farmers 农场主 74—75
    attachment to land 对土地的归属感 79
    business objectives 商业目标 79
    market and political power 市场和政治力量 78
farming, image of 农业的形象 87
farming systems 耕作制度 84—85, 92, 99
farrowing crates 产仔箱 99
fat metabolism in animals 动物的脂肪代谢 35
fats and oils 脂肪和油类 2, 18, 35
feeding farm animals 饲养农场动物，

见 animals, animal nutrition
fertilizer 化肥 2, 14—16, 23, 81—82, 90—91
　compound fertilizers 复合肥 15
　diminishing response to 递减反应 16
fibre crops 纤维作物 54
Food and Agriculture Organization (FAO) of the United Nations 联合国粮食及农业组织 68, 70—71, 91, 94, 100
fowl pest 鸡瘟 3
fungi 真菌 2, 6

G

geese 鹅, 见 poultry
genetic improvement 遗传改良 41—44
genetically modified organisms 转基因生物 109—113
gestation periods 妊娠期 43
glasshouses 温室 18, 23, 70
global warming 全球变暖 94—95, 105—109
glyphosate herbicide 草甘膦除草剂 111, 113
GMOs 转基因生物, 见 genetically modified organisms
goats 山羊 33, 35, 43, 58
government agricultural policy 政府的农业政策, 见 agricultural policy
grains 谷物, 见各作物
grass 牧草 27, 36—37, 54
greenhouse gases 温室气体 82—84, 106—108

greenhouses 温室, 见 glasshouses
guineafowl 珠鸡, 见 poultry
guinea pigs 豚鼠 32

H

hay 干草 30, 36, 96
hedgerow loss 树篱损毁 96
Heller, Joseph 约瑟夫·海勒 103
hens 母鸡, 见 poultry
herbicides 除草剂 25—26, 29, 94, 111, 113
heritability 遗传性 42
honeybees 蜜蜂 3
horses 马 33, 36, 41, 58
housing farm animals 安置农场动物 44—45, 98
hydroponics 水培 70

I

insecticides 杀虫剂 3, 26, 93, 96
insect pests 虫害, 见 animals, animal health and disease; crops, crop pests
International Panel on Climate Change (IPCC) 政府间气候变化专门委员会 105, 109
International Rice Research Institute 国际水稻研究所 26
international trade in farm products 农产品的国际贸易 64—68, 102, 104
　exports 出口 65—66, 102
　imports 进口 64—65, 102
irrigation 灌溉 9, 23—24, 71—72, 107

J

Jeffersonian Agrarianism 杰斐逊式农论 88

L

labour 劳动力 73—80
lactation 哺乳 37—38, 42
land 土地 70—73, 115
    arable land 耕地 70
    land area, world 世界土地面积 70
    land productivity 土地生产率 71, 116
    loss, as a result of climate change 气候变化导致的损失 107
landscape, impact of agriculture on 农业对景观的影响 95—97, 101
legumes 荚果 27, 30
lentils 兵豆 27, 56
lime (soil conditioner) 石灰（土壤改良剂）14
livestock 畜禽；参见 animals; individual species
    livestock feed 畜禽饲料 51, 57, 59
    livestock products 畜禽产品 58
llamas 美洲驼 32

M

Maasai pastoralists 马赛牧民 79
maize 玉米 2, 17, 28, 30, 55, 57, 59, 66；参见 corn
manure 粪肥 2, 13—15, 82
meadow 草地 70

meat industry 肉类加工业 53—54
meat production, world 世界肉类生产 57—58
metabolic pathways 代谢途径 18
milk 奶 33, 34, 38, 66
    milk production, world 世界奶产量 58
    milking 产奶 48—49
millet 谷子 16, 27, 55
mules (sheep breed) 骡羊（绵羊品种）39
mules (equine) 骡子（马）33
muskoxen 麝牛 33
mustard 芥菜 27

N

Nash, Roderick 罗德里克·纳什 95
National Farmers' Union (NFU) 全国农民联盟 103
National Grange (USA) 国家农庄（美国）103
Nitrate Vulnerable Zones 硝酸盐脆弱区 93
non-farm rural economy 非农业农村经济 78
non-ruminant animals 单胃动物 34, 36
notifiable diseases 依法须报告的疾病 47

O

oats 燕麦 27, 55
Ogallala aquifer (USA) 奥加拉拉含

水层（美国）72
oilseed rape 油菜，见 rape
organic farming 有机耕作 92
ostriches 鸵鸟 33

P

pasture 牧场 11, 70
peas 豌豆 12, 27
peasants 农民 75
performance testing 性能测试 41
pesticides 农药 26, 82, 91, 96
    pesticide resistance 抗药性 93, 113
pests 害虫，见 animals, animal health and disease; crops, crop pests
photosynthesis 光合作用 17—20
pigs 猪 33—34, 36, 58, 70
plant breeding 植物育种 26—27, 109—113
ploughs and ploughing 犁与犁地 28
pollution 污染 93—94
population, world 世界人口 113—114
potatoes 马铃薯 2, 9, 16, 22—23, 55, 66
poultry 家禽 1—2, 32, 36, 58, 66
precision farming 精耕细作 16, 93—94
pregnancy 怀孕 38
progeny testing 后代测试 41—42
proteins 蛋白质 2, 18
    protein metabolism in animals 动物的蛋白质代谢 35
pulses 豆类 30, 56；参见 legumes

R

rainforest 雨林 94—95

rape 油菜 27, 30, 56, 112
reindeer 驯鹿 33
Ricardo, David 大卫·李嘉图 70
rice 水稻 11, 17, 27, 30, 55, 59, 66, 111
root crops 块根作物 2, 55
rough grazing 粗放式放牧 54
ruminant animals 反刍动物 34—37
rye 黑麦 27, 55

S

sheep 绵羊 32, 58, 66
sheep breeds 绵羊品种 40
silage 青贮饲料 30, 57
silk moths 蚕 32
soil 土壤 4—12
    acidity 酸度 13—14
    alluvium 冲积物 5
    clay 黏土 7—10
    classification 分类 11—12
    composition 组成 8
    fauna 动物区系 6
    fertility 肥力 7—8
    formation 形成 5—7
    glacial drift soils 冰碛土 5
    loam 壤土 8
    loess 黄土 5
    structure 结构 9—11
    texture 质地 7—8
    tropical and semi-tropical 热带与亚热带 11
sorghum 高粱 27, 55
soya beans 大豆 2, 30, 55, 57, 66, 111, 112
staple foods 主食 59, 61

starch 淀粉 18

subsistence farmers 勉强维持生活的农民 64

sugar 糖 66

sugar beet 甜菜 9, 29, 55

sugar cane 甘蔗 27, 55

Svalbard Global Seed Bank 斯瓦尔巴全球种子库 19

swedes 芜菁甘蓝 27

T

tea 茶 56, 66, 67

tilth 耕性 10

tissue growth curves 组织生长曲线 38

tractors 拖拉机 81, 82, 87; 参见 agricultural tools and machinery

turnips 芜菁 30

U

United States of America (US) 美国 2, 30, 66, 103

V

vaccines 疫苗 46, 49

vegetables 蔬菜 56

vernalization 春化作用 19

veterinary medicines 兽药 46

vitamins 维生素 2

W

water footprint 水足迹 72—73

weeds 杂草 3, 20—21, 25

   weed control 杂草防治 25; 参见 herbicides

wheat 小麦 2, 28, 54, 55, 59, 66

wildlife, impact of agriculture on 农业对野生动物的影响 94—97, 101

World Commission on Environment and Development 世界环境与发展委员会 89

world food balance 世界粮食平衡 113—115

World Trade Organisation 世界贸易组织 67

wool 羊毛 33, 58

Y

yaks 牦牛 32

yams 薯蓣 2, 16, 55

yields 产量 54, 55—56, 115—116

   effects of climate change 气候变化的影响 108—109

   of eggs 蛋的产量 1

   of wheat, world record 小麦产量的世界纪录 54

Z

zebu cattle 瘤牛 39

zoonoses 人畜共患病 47

Paul Brassley and Richard Soffe

AGRICULTURE

A Very Short Introduction

Contents

Acknowledgements i

List of illustrations iii

List of tables v

Introduction 1

1 Soils and crops 4

2 Farm animals 32

3 Agricultural products and trade 51

4 Inputs into agriculture 69

5 Modern and traditional farming 86

6 Farming futures 100

Further reading 117

Acknowledgements

We have spent our working lives thinking and talking about agriculture with numerous acquaintances, students, colleagues, and friends, all of whom have affected our views in one way or another. For their helpful comments on the manuscript of this book we are especially grateful to Angela Brassley, Caroline Brassley, Charles Brassley, Charlie Clutterbuck, Alan Cooper, Steve Jarvis, Anita Jellings, Rob Parkinson, Alison Samuel, our editors at Oxford University Press, Andrea Keegan and Jenny Nugee, and our anonymous readers. For all her work on the figures and photographs, we are most grateful to Carrie Hickman, and we would like to thank Dan Harding for his careful, sensitive, and expert copy editing.

List of illustrations

1 The US Dust Bowl in the 1930s **5**
   © Everett Historical/Shutterstock.com

2 Soil structure in detail **10**
   Soil structure formation diagram (Fig. 1.3) from R. J. Soffe (ed.) *The Agricultural Notebook*, 20th edn, Wiley. © 2003 by Blackwell Science Ltd, a Blackwell Publishing company

3 The most important food grains: (a) wheat, (b) millet, (c) rice, and (d) maize **17**
   (a) © marima/Shutterstock.com;
   (b) © huyangshu/Shutterstock.com;
   (c) © kuma/Shutterstock.com;
   (d) © Maks Narodenko/Shutterstock.com

4 Centre pivot irrigation promotes crop growth in dry California **24**
   © Craig Aurness/Corbis

5 Relative growth rates of an animal's body tissues from conception to maturity **38**
   Growth curves diagram (Fig. 16.36) from R. J. Soffe (ed.) *The Agricultural Notebook*, 20th edn, Wiley. © 2003 by Blackwell Science Ltd, a Blackwell Publishing company

6 Farmers in Somoa village, upper west Ghana, sell their farm products, including millet, at market **74**
   TREE AID, photograph by Carrie Brassley

7 Crop land displacing the Amazon rain forest **95**
   © age fotostock/Alamy

List of tables

1 Particle sizes and properties of soil minerals **8**
Source: R. J. Parkinson, 'Soil Management and Crop Nutrition', in R. J. Soffe (ed.), *The Agricultural Notebook*, 20th edn, p. 4. © 2003 by Blackwell Science Ltd, a Blackwell Publishing company

2 World farm animal population 2012 **33**
FAO.FAOSTAT. Live Animals: Number of Heads (FAO 2015) Accessed 29 July 2014 <http://faostat3.fao.org/faostat-gateway/go/to/browse/Q/QA/E>

3 Approximate gestation periods **43**

4 The principal crops **55**
FAO.FAOSTAT. Production quantities (FAO2015) Accessed 26/7 April 2014 <http://faostat3.fao.org/browse/Q/QC/E>

5 Major livestock products, world total 2012 **58**
FAO.FAOSTAT. Livestock primary (FAO2015) Accessed 9 October 2014 <http://faostat3.fao.org/faostat-gateway/go/to/browse/Q/QL/E>

6 Income effects on the demand for food **62**
A. Regmi, M. S. Deepak, J. L. Seale Jr, and J. Bernstein, Cross Country Analysis of Food Consumption Patterns, Economic Research Service, US Department of Agriculture., tables B-1 and B-2, available at <http://www.ers.usda.gov/media/293593/wrs011d_1_.pdf> (accessed 20 October 2014)

7 The top ten importers and exporters of agricultural products in 2012 **65**
World Trade Organization, *International Trade Statistics 2013, part II, Merchandise Trade*, © World Trade Organization, 2013, <http://www.wto.org/english/res_e/statis_e/its2013_e/its13_merch_trade_product_e.pdf, table II.15> (accessed October 2014)

8 World exports as a percentage of world production in 2011–12 **66**
FAO.FAOSTAT. Crop and livestock products. Exports of selected countries (FAO 2015) Accessed 10 October 2014 <http://faostat3.fao.org/browse/T/TP/E>

9 Water footprint per unit of nutrition **73**

Data from M. M. Mekonnen and A. Y. Hoekstra, 'A Global Assessment of the Water Footprint of Farm Animal Products', *Ecosystems*, 15, 2012, pp. 401–15

10 The distribution of farms and land in the late 20th century **77**

Data from *Food and Agriculture Organization of the United Nations*, 2014, S. K. Lowder, J. Skoet, and S. Singh, 'What Do We Really Know about the Number and Distribution of Farms and Family Farms Worldwide?', background paper for *The State of Food and Agriculture 2014*, ESA Working Paper No. 14–02, Rome: FAO <http://www.fao.org/docrep/019/i3729e/i3729e.pdf>. Reproduced with permission

11 Purchased inputs in agriculture in various countries, 2000–10 **82**

FAO, *FAO Statistical Yearbook 2013: World Food and Agriculture*, Rome: FAO, 2013. Reproduced with permission

12 UK greenhouse gas emissions **83**

Source: Data from P. Scarborough, P. N. Appleby, A. Mizdrak, A. D. M. Briggs, R. C. Travis, K. E. Bradbury, and T. J. Key, 'Dietary Greenhouse Gas Emissions of Meat-Eaters, Fish-Eaters, Vegetarians and Vegans in the UK', Climate Change 125, 2014, pp. 179–92

Introduction

Consider an egg. Agriculture is about the getting a hen to lay the egg. It's also about producing the raw materials for other things that you might eat with it, such as bread and butter.

Where did the hen come from? Of course it came from another egg, but in this case one that was fertilized by a cockerel so that a chick developed inside it. Most commercial-scale egg producers in industrialized countries now buy their laying birds from specialist breeders who employ geneticists to develop more productive hens, each of which can now lay 300 eggs or more per year. This is roughly twice their output in 1950.

Having obtained his hen, the farmer then has to find somewhere for her to live. Hens are the domesticated descendants of Indian jungle fowl, and left to themselves they retain some of the behaviour of their wild ancestors, which would find somewhere secluded to lay their eggs, and roost in trees at night. People who keep hens in small numbers therefore often have hen houses with laying boxes containing nesting material and perches on which the hens can roost. At the opposite extreme, conventional battery cages used to have two or three hens standing in a small cage on a wire floor. This type of cage has been banned in the European Union (EU) since 2012. It has been replaced by 'furnished' cages which are much bigger, contain more hens, often between forty and eighty, and have nest boxes and dust-bathing areas. About half of the eggs produced in the UK now come from birds in cages, with the rest mostly produced by free-range flocks. In the US over 90 per cent of eggs are produced in battery cages, although one or two states—California, for example—have banned their use.

The farmer also has to ensure that the hen has food and water. Hens are natural scavengers. They will eat seeds, fruits, grass and other green leaves, worms, and insects, and they pick up small pieces of grit which go into their gizzards and help physically to break down their food. When the hens are kept in cages, and also in many free-range systems, the farmer has to provide all the nutrients that the hen will transform into eggs: carbohydrates, proteins, fats, minerals (especially calcium for the shells), and vitamins. All of these are combined into pellets or crumbs in a mill, which on large poultry operations may be on the farm, but is more often on a factory site, sometimes close to a seaport for ease of access to imported ingredients. The carbohydrates mostly come from cereals such as wheat, barley, or maize (known as 'corn' in the US), which have been grown in fields, some of which may be thousands of miles away. Similarly, soya beans, many of which are produced in Latin America, are one of the main sources of proteins and fats for European and North American hens. Thus the egg, like many animal products, is linked to arable farming, or the production of crops.

Arable farmers produce the cereals, protein and oil crops, root crops, such as potatoes or (in tropical countries) cassava and yams, and all the other plants and plant-derived raw materials that make up animal food and the great bulk of human food. To grow cereals, for example, they take the seed, which in industrialized and many developing countries is now a very sophisticated science-derived source of genetic material, and sow it into soil which has been prepared with the aid of powerful machinery and enriched by manures and fertilizers. The growing plant can be protected from attacks by fungi and insects, and from competition by weeds, by the enormous variety of herbicides, fungicides, and insecticides that have emerged from the chemical industry over the last seventy years. Finally the crop is harvested by large machines which separate out the grain that will be used to feed people and chickens from the rest of the plant.

But while all that is happening in developed countries and in some countries of the South, at the same time there are still many countries in which the seedbed is prepared by farmers bending from the waist and using the same kind of hoe that their ancestors have been using for hundreds of years. These farmers might not be able to keep hens because a disease called fowl pest is endemic to their area and they cannot afford the few pence that it would cost to vaccinate them against it.

The agribusiness corporation producing corn and soya beans using enormous machines fifty miles south of Chicago, the woman with her hoe and her plot of cassava in Mozambique, the Chinese collective farm worker in the rice fields, and the German family with their part-time dairy farm and their day jobs in Munich are all engaged in agriculture. This *Very Short Introduction* sets out to identify the common features of their activities and the universally applicable principles that determine what they do, to explain why the differences between them exist, and to explore some of the controversies that arise from their activities. It is neither a criticism, nor a defence, of either science-based farming methods or traditional farming practices. It recognizes the diversity of a world in which developed and developing agriculture can be found anywhere and everywhere.

Chapter 1
Soils and crops

Soil

In May 1934 a wind storm lifted a black cloud of soil from the Great Plains and dumped twelve million tonnes of it on to the city of Chicago (see Figure 1). Two days later the storm reached the eastern seaboard and the dust drifted in through the windows of the White House in Washington. It was a salutary reminder that human life is dependent upon just a few inches of fragile topsoil, as important to farmers as sunlight. As a resource to anchor and feed the crops, it is a living medium containing thousands of living organisms. Across the world the type and depth of soil is fundamental to the crops that can be grown. The first part of this chapter looks at what soils are and how they differ, how they are classified, and how farmers manage them.

Soil is a complex mixture of air, water, minerals, and organic matter that has evolved, in many cases, over thousands of years. Typically, air may account for 25 per cent of the volume of a well-managed topsoil, and water another 25 per cent. The organic matter could take up about 5 per cent of the volume, so that the actual mineral fraction, varying from large stones to tiny clay particles, only comprises 45 per cent of the volume, although much more of the weight of the soil.

1. The US Dust Bowl in the 1930s: a vivid image of wind-blown soil on the move.

The type of soil found in any particular location depends upon the parent material, the climate, the topography of the site, the various organisms living in and on the soil, and time. Some parent materials are solid rocks such as granite, sandstone, chalk and limestone, or slate, whereas others are superficial deposits such as riverine alluvium, which is material that has been carried by a river and then deposited on its flood plain. Alluvial soils can be some of the most fertile in the world, whereas soils formed on sand dunes are often at the opposite end of the fertility spectrum. Loess, like sand dunes, is formed of material transported by the wind, but in this case involves much finer material. It covers extensive areas of North America and northern China, in the latter case being derived from the Gobi Desert. There are also peat soils, formed in wet conditions, glacial drift soils, formed on materials deposited by melting glaciers, marine clay soils, such as those in parts of the Netherlands, derived from material originally laid down under the sea, and volcanic ash soils.

What develops from these and other parent materials then depends upon the local climate, in particular the prevailing temperature and moisture, the organisms that come to live there, and the topography of the site, especially the slope. The greater the slope, the easier it is for repeated freezing and thawing, or torrential rain, to move material downhill, so gently sloping sites tend to have deeper soils than steep slopes. Increasing temperatures accelerate the weathering of minerals and the breakdown of organic matter, and the greater the soil moisture the more soluble elements such as calcium move down from the surface layers, so that wet tropical or semi-tropical soils are often quite acid whereas drier soils in areas of similar temperature are less so. The plants growing on the soil extract water and nutrients from it; when they die, under natural conditions, they become part of the decomposing surface litter, and following their breakdown the nutrients are again available for reabsorption.

In tropical forests the amount of litter deposited on the surface may be up to ten times that found in a northern pine forest, but such is the rate of decomposition in the warmer conditions that tropical soils may still have only low levels of organic matter. Some of the decomposition and mixing into the soil of plant debris is carried out by small animals such as earthworms, millipedes, springtails, mites, nematodes (eelworms), ants, and termites. The other decomposers are the fungi, bacteria, actinomycetes (which are unicellular like bacteria but produce mycelial threads like fungi), algae, and viruses. At the opposite end of the scale are the vertebrates such as rabbits, moles, prairie dogs, and the blind mole rat, all of which contribute to the mixing of soil layers by their burrowing activities. And of course farmers mix the soil when they cultivate it.

There is a cyclic process at work here, generally known as the carbon cycle, in which atmospheric carbon dioxide is fixed into plants (see the following section 'How plants grow'), which either die and are decomposed, or are consumed by humans and other animals and so partly fixed into their bodies, with the residue emerging as dung and urine to be decomposed in their turn and become part of the soil organic matter, some of which will be oxidized and so return to the reservoir of atmospheric carbon dioxide. The rate at which all this happens has clearly been affected by the spread of agriculture, because grazing animals remove growing plants and convert them both to dung and urine, but also to gaseous carbon dioxide and methane. It has also been affected, probably to a greater degree, by the exploitation of fossilized carbon in the form of coal and oil, which adds to atmospheric carbon dioxide levels.

All these factors contribute to the formation of different types of soil, as we shall see later, but for the individual farmer what probably matters most is the texture and structure of the soil. Texture is determined by the proportion of mineral particles of different sizes in the soil, as shown in Table 1, which shows the classification system employed in Britain.

Different soils will contain more or less of these various particles. One made up of 40 per cent sand, 40 per cent silt, and 20 per cent clay, for example, would be classified as a loam according to the US Department of Agriculture system, whereas one with equal quantities of all three would be a clay loam. Accurate determination of particle size distribution requires a complex process of chemical treatment, sieving, and sedimentation, but experienced soil scientists can make an effective enough assessment by a process known as 'spit and rub'. The soil is moistened between the fingers and thumb, and the feel of the sample gives an idea of the texture. Coarse sand feels gritty, fine sand slightly less so, silt makes the sample silky or slippery, and clay makes the sample sticky.

The texture of a soil affects how easy it is to work and how fertile it is. A very sandy soil will drain easily, but will also be subject to drought and not especially fertile. Those farming such soils often

Table 1 Particle sizes and properties of soil minerals

	Diameter (mm)	Characteristics
Gravel or stones	Greater than 2	
Coarse sand	0.2–2	Beach sand, visible grains
Fine sand	0.06–0.2	Egg timer sand
Silt	0.002–0.06	Flour, only visible with a hand lens
Clay	Less than 0.002	Putty, only visible through an electron microscope

Source: R. J. Parkinson, 'Soil Management and Crop Nutrition', in R. J. Soffe (ed.), *The Agricultural Notebook*, 20th edn, p. 4. © 2003 by Blackwell Science Ltd, a Blackwell Publishing company.

say that they need a shower of rain every day and a shower of manure on Sundays, although they may not use those precise words. Clay soils, on the other hand, are much more fertile but much more difficult to work.

What is the reason for the difference? Sand and silt-sized soil particles have a relatively small surface area per gramme, and are largely composed of quartz (silicon dioxide) particles, which are chemically unreactive. Consequently they are less likely to retain water by capillary attraction and nutrients in chemical combination. Clay particles, on the other hand, have an enormous surface area for a given weight, often a thousand times greater than that of sand, and are chemically reactive, so they retain water and nutrients. Thus they are potentially more fertile, but on the other hand it requires more power to pull a plough or any other cultivator through a clay soil than a sandy soil, and clays are slow to drain and difficult to rewet in times of drought. Clay soils are often therefore described as 'cold' and 'heavy', whereas sandy soils are said to be 'light'. Most soils, of course, and certainly the best ones from a farming viewpoint, such as loams or clay loams, have neither too much sand nor too much clay and so are both fertile and easy to work.

Farmers can affect the water content of the soil, and so improve its workability, lengthen the growing season, improve the efficiency of fertilizer use, and encourage crops to root more deeply, by artificial drainage. The impact on crop yields in Britain has been found to be between 10 and 25 per cent. Various techniques are employed, from open ditches leading to streams and rivers, to tile or pipe drains and mole drains. The mole, in this case, is a pointed cylinder of about 70 mm diameter attached to the base of a vertical blade which is pulled through the soil by a powerful tractor. In clay soils the channels so produced may last for up to five years. Tile or pipe drains, laid by specialist machines, last far longer but are correspondingly more expensive to install.

The opposite process, of course, is irrigation, which has been carried out in some countries for hundreds if not thousands of years: the seasonal inundation of the Nile Delta is an obvious example. In temperate countries it is often only higher value crops such as potatoes, sugar beet, and vegetables that will repay the cost of irrigation, but in tropical areas irrigation is an integral part of the cultivation of paddy rice crops. One estimate suggests that 20 per cent of the global area of arable land, accounting for about 40 per cent of total crop production, is now irrigated. Much of this land is accounted for by the Asian rice crop, but there is a lot in the US too.

The other important consideration for the farmer is the structure of the soil, because that, too, affects how easy it is to work. The development of a soil, as opposed to a simple mixture of sand, silt, and clay, requires the mixture of the mineral particles discussed earlier with organic matter, by the activity of soil organisms over time. The result is the formation of stable aggregates that contain the water and air needed by plant roots and other soil organisms (see Figure 2).

The pore spaces can be filled with either water or air or a mixture of the two, and a well-structured soil has an even

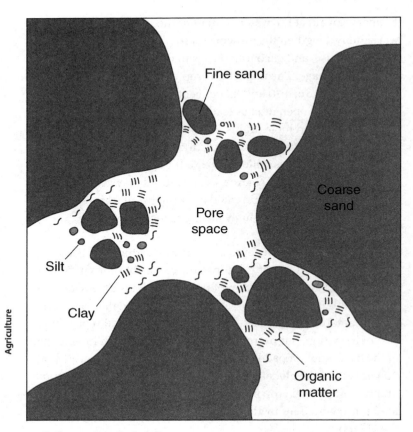

2. Soil structure in detail.

granular topsoil with gradually larger aggregates at greater depth. It is relatively easy, in such soils, for cultivation implements to produce the finely structured surface soil that farmers call 'a good tilth', which provides the ideal conditions in which plants can germinate and grow. In contrast, in a poorly structured soil, such as a clay loam that has been worked when too wet, the clay will have become 'puddled', or brought together in large lumps, often with an impervious pan at 25–30 cm depth which impedes drainage.

Farmers cannot change the texture (percentages of sand, silt, and clay) of their soil, but their farm management decisions can affect its structure. Continuous cropping—or, in the wrong circumstances, even regular arable cropping—can reduce the level of organic matter in the soil, making such structural problems more likely, whereas the use of permanent pasture, or at least long periods under grass, and regular additions of manure and lime, all help to improve soil organic matter and soil structure. In extreme circumstances the structure deteriorates to such an extent that it becomes susceptible to wind erosion, as was seen in East Anglia in the 1960s and most famously in the Dust Bowl of the Midwestern United States (see Figure 1).

All these variations in parent material, texture, structure, organic matter, and topography mean that there is an immense variety of soils in the world, for which soil scientists have developed various systems of classification, many of them involving extremely complex descriptive terms. The most widely used systems are related to climatic variables, mainly rainfall and temperature, and consequent vegetational zones.

Thus brown earths (sometimes called Cambisols) develop under temperate deciduous forests and make good mixed farming soils, and chernozems are the black soils that develop on steppes or prairies and form the arable farming soils of the US Midwest and the Ukraine. Acid, leached podzols develop in the cold and wet conditions found in coniferous forests, heaths, and moorlands, whereas in the warm wet conditions of the tropics, highly weathered lateritic soils (also called latosols or ferralsols), sometimes red or yellow in colour, are found. They are acid and nutrient deficient, and although they can support luxuriant forest sustained by rapid nutrient recycling at the surface, they form only poor agricultural soils. Gleys are waterlogged soils which in temperate regions need drainage and careful management, but in the tropics constitute some of the major rice-growing soils.

These are just a few of the wide range of soils recognized in climate-based classification systems, but there are other approaches which place greater emphasis on the significance of parent material and topography, and thus develop global classifications relating to geography and land form.

The soil, therefore, physically supports growing crops and supplies water to their roots. Dissolved in the water are the various nutrients that plants need to help them extract their main constituent, carbon, from the carbon dioxide in the air, and we shall look further at those in the following section.

Crop nutrients

Plants get their major constituent elements—carbon, hydrogen, and oxygen—from the air they take in through their leaves and the water taken up by their roots. Apart from those, they require three elements in large quantities (macronutrients), another three in rather smaller amounts, and very small amounts of another eight, usually called the micronutrients.

The three major nutrients are nitrogen (N), phosphorus (P), and potassium (K). Most of the nitrogen is acquired from the soil water in the form of nitrate ions, and is a constituent of the plant's proteins and chlorophyll. Without it plants become yellow and fail to thrive. Ultimately, the nitrogen in plants comes from the air, but first it has to get into the soil. A small amount arrives dissolved in rainwater, but more comes from the activities of the soil-living algae or bacteria, such as *Azotobacter*, or from the nitrogen-fixing nodules on the roots of leguminous plants such as peas, beans, clovers, and even gorse. From these organic sources it is then converted by further bacterial activity to nitrate ions, which are very soluble in water, and in this form are taken into the plant roots. They may also, however, be leached out as water drains away, or converted into atmospheric nitrogen again, or taken up by soil organisms to once again become part of the organic reserve in the soil. What most manures and mineral fertilizers do is add to the amount available at the nitrate stage of the cycle.

Phosphorus is required for the energy and protein metabolism of plants, and without it roots do not grow well. There is usually plenty of it in the soil, but in a relatively insoluble form, so in some circumstances only the small amounts in solution may be available to plants. Potassium is involved in the water control and transport mechanisms of plants, so growth rates are reduced without it. It is released during the weathering of clays in the soil, and can be lost in drainage water as a result of its solubility. In normal circumstances and most farming systems, it is only these three major nutrients that are lost in sufficient quantities to require replacement by fertilizers.

There are, however, other elements that may sometimes need to be replaced in more intensive farm management systems. Sulphur is a constituent of some amino acids and enzymes, and in normal circumstances there are sufficient soil reserves, but since the decline of sulphur emissions from coal-fired power stations in Europe there have been signs of sulphur deficiencies in some areas. The other two elements in this category are calcium and magnesium, which play a part in controlling the acidity of the soil (see next paragraph). Finally, there are the micronutrients: boron, copper, zinc, manganese, molybdenum, iron, chlorine, and cobalt, many of which play important roles in plant enzymes but are usually required in such small quantities that most soils contain adequate amounts. For example, whereas a cereal crop producing 6 tonnes of grain and 3.5 tonnes of straw would remove 120 kilogrammes of nitrogen from a hectare of land, it would result in the loss of less than a kilogramme of zinc.

Soil acidity also affects the availability of nutrients. It is measured by the pH of the soil, a pH of 7 being neither acid nor alkaline, with the acidity increasing as the pH decreases. A cultivated soil normally has a pH between about 5 and 7, although on calcareous soils such as those on limestone and chalk the value could be between 7 and 8. The most acid upland peat soils might fall as low as 3.5. At this level of acidity the availability of all of the macronutrients and some of the micronutrients is reduced. In a slightly acid soil, with a pH between 6 and 7, nutrient availability tends to be at a maximum, as is micro-organism activity, so farmers recognize the need to maintain it at that level, by adding appropriate dressings of liming materials such as ground limestone. Lime, which also aids good soil structure, should therefore be seen as a soil conditioner rather than a fertilizer.

Manures and fertilizers

From the preceding discussion, it is clear that the purpose of manures and fertilizers is to replace the plant nutrients that cropping and grazing remove, so they are mostly valued for the N, P, and K that they contain. There is no hard line between what constitutes a manure and what is a fertilizer. Essentially, manures come from organic sources—the classic example is farmyard manure, which is a mixture of straw, dung, and urine—and the feedstocks for fertilizers are produced in chemical factories (in the case of N fertilizers) or mined (in the case of P and K). But there are plenty of examples of products which are midway between these two extremes, such as blood and bone meal, or dried chicken droppings, which originate from an organic source but have then gone through a manufacturing process.

The nutrient content of manure is relatively low: even the most concentrated, which is poultry manure, only contains about 2 per cent of N and P and 1 per cent of K. The additional benefit of farmyard manure is that it contains partially decomposed organic matter which helps to improve the structure of the soil. In recent years, on highly mechanized farms, animal dung and urine have been collected as a liquid slurry, which is then applied directly to the land, without the addition of straw, so this will not produce the same soil structure benefits.

Organic fertilizers (as opposed to manures) are more concentrated, with hoof and horn meal and dried blood containing 12–14 per cent N. The range of such materials used in the past was enormous: guano, gas lime, fish, blubber, 'greaves' (candle-makers' waste), furriers' clippings, feathers, wool, linen rags, shoddy, fellmongers' poake (carcass remains), and rape dust all appeared on a list compiled in 1860. Similarly, many materials, from burnt lime to marl and shell sand, have been used in the past for their lime content.

Inorganic fertilizers are the most concentrated, with available nutrients accounting for up to half the weight of the material applied. There are several kinds of nitrogenous fertilizers, such as ammonium nitrate, urea, and anhydrous ammonia, but in each case the nitrogen content has been fixed from the atmosphere (a process which requires a lot of hydrocarbon fuel, now often in the form of gas) before being combined with other elements. Similarly, most phosphate fertilizers are derived from rock phosphate, which can be used directly as a fertilizer, but is very slow to release its phosphorus. Consequently it is usual to treat the rock phosphate with sulphuric acid (to make superphosphate) or phosphoric acid (to make triple superphosphate, with a higher P content).

Potassium is mined in the form of potassium chloride, which is either used directly or treated with sulphuric acid to make potassium sulphate. All these individual nutrients can be used on their own—in which case they are often called 'straights'—but they may also be combined to make 'compound' fertilizers which contain a mixture of nutrients, sometimes formulated to suit the requirements of particular crops. Thus a good general purpose compound fertilizer would be a 20;10;10, containing 20 per cent N and 10 per cent each of P and K, whereas a fertilizer designed to replace the nutrients removed by a cut of silage would require no P, so might be a 25;0;15 (i.e. 25 per cent N and 15 per cent K). How much of any of these to apply to a specific field or crop depends upon a variety of considerations, from the existing N, P, and K status of the soil to the response of the crop, its sale value, and the cost of the fertilizer. Most crops have a diminishing response curve, meaning that the additional crop yield response decreases as the level of fertilizer application increases.

The sustained production of crops and grass over long periods of time therefore requires healthy, well-managed soils. Some farmers believe that these can best be produced by avoiding the use of mineral fertilizers and relying on organic methods to recycle nutrients. Others use precision farming methods to maintain soil fertility by placing mineral fertilizers only where they are needed. Farmers have to make the right management decisions according to the resources available to them and the crops they wish to grow.

Crops

Humans live on plants, either directly, by eating them, or indirectly, by eating the flesh or products, such as milk, that animals produce as a result of eating them. A few of the plants, such as blackberries or mongongo nuts, might be gathered from the wild, but most are cultivated, in which case they are called crops. Some sort of starchy energy crop forms the basis of human diets almost everywhere, often a grain crop, such as wheat, millet, rice, or maize (see Figure 3). Where conditions are unsuitable for these, roots and tubers such as cassava, yams, or potatoes are grown. The following sections of the chapter look at how plants grow, what prevents them from growing, and how we can improve their growth, before moving on to some of the major crops and cultivation systems.

3. The most important food grains: (a) wheat, (b) millet, (c) rice, (d) maize.

How plants grow

How do farmers capture the energy in sunlight and make it available so that you, the reader, have the energy to turn the pages of this book? They employ plants containing chlorophyll, which is the stuff that makes plants green. It absorbs radiation which provides the energy to drive a series of biochemical reactions within the leaves which take carbon dioxide (CO_2) from the air and combine it with water taken up by the roots to make simple carbohydrate molecules—sugars—and release oxygen (O_2). The process is called photosynthesis and is summarized in the equation

$$6CO_2 + 6\,H_2O \xrightarrow{\text{light}} C_6H_{12}O_6 + 6O_2$$

and it is difficult to exaggerate its importance. Without it there would be no life as we know it on this planet. It happens in all green plants, from the shortest grass to the tallest trees, and it is what produces 95 per cent of their dry matter (i.e. the part that isn't water). The simple sugars such as glucose can combine together in long chains (called polymers) to form starch, which is the form in which most plants store their energy, and cellulose, which is one of the main constituents of the plant cell walls. Fats and oils can also be produced by further transformations of carbohydrates, and, with the addition of nitrogen and sometimes other elements, proteins too.

Clearly, therefore, the speed at which the crop grows depends upon the rate of photosynthesis, and that in turn depends upon light, temperature, the carbon dioxide concentration and the availability of water and nutrients. As the temperature and light intensity increase so does the rate of photosynthesis, as long as water is available. For most temperate crops, which use what is called the C_3 metabolic pathway for carbon fixation, the limiting factor is the CO_2 concentration in the atmosphere. This is why increasing levels of this gas due to the burning of fossil fuels have a fertilizing effect. It is also why some growers enhance the CO_2 levels in glasshouses.

Many tropical crops such as maize, sugar cane, millet, and sorghum use the alternative C_4 metabolic pathway, in which photosynthesis is not limited by the CO_2 concentration until much higher light intensities. Plants with this metabolism can have high photosynthetic rates in hotter drier conditions, so research is currently going on to identify the genes responsible and to use them to produce a C_4 variety of rice. It has been claimed that this could yield up to 50 per cent more grain than current C_3 varieties.

As a plant grows, the dry matter or total biomass increases because the number of cells increases. The plant progresses from seed germination through vegetative development, reproductive development, and seed development by means of cell specialization (differentiation) and organization. Depending on its life cycle (annual, biennial, or perennial) and the part of the plant that is useful and harvested as food, a crop may reach senescence (death) after weeks, months, or years.

Before germination most seeds are resistant to cold and drought stress and can often survive for long periods. Those in the Svalbard Global Seed Bank, for example, are kept in a disused mine at below-zero temperatures in the permafrost. Most temperate crops need a temperature of at least 1–5°C before they will germinate at all, with optimum germination at between 20–25°C, but the optimum for rice is 30–35°C. In the vegetative stage the growing point(s) (meristems) of the plant produce leaf and stem material, and in the reproductive stage flowers, which, following pollination and fertilization, produce seeds. For grain crops, farmers only want enough stem and leaf development to maximize the photosynthetic (green) area of the plant, because it is the seed that they will harvest and sell, whereas they do not want potatoes to produce seed at all, because it is the tubers that are harvested as food.

Most crops grow faster as the temperature increases, and crops such as maize that originated in warmer climates normally require higher temperatures for optimum growth than those such as wheat or ryegrass that were originally found in cooler parts of the world. Temperature drives the rate of development, and often controls the timing of the change from one stage of growth to another. For example, some varieties of wheat are designed to be sown in the autumn (winter wheat), and require a period of cold weather (vernalization) in order to set seed in the following summer. Some plants also respond to day length. Some rice varieties, for example, will not flower until the days begin to get shorter, whereas most temperate plants are 'long-day' types, needing increasing day length to begin flowering, which ensures that they do so in the summer.

All these are features of the individual plants, but the crop is a collection of plants, and the interactions of the individuals are also important. The farmer seeks to manage the crop as a whole to maximize the amount of light intercepted at the time when the days are longest and the sun is at its most powerful. Thus if it is sown too late it may not have produced enough leaves to cover the ground, and sowing too thinly will have the same result. A good rich soil and enough water will also encourage leaf growth. On the other hand, take any of these factors too far and the result will be a decrease in yield, as higher leaves shade out the lower ones, individual plants have insufficient room in which to grow, or rapid growth or water shortages produce weak or floppy stems that do not remain erect.

A balance has to be struck between the yield of the individual plant and of the crop as a whole. The skill and experience of the farmer and the work of the agricultural scientist are needed to determine optimum seed and fertilizer rates and sowing times in order to maximize the output of the crop.

What prevents plants from growing?

It follows from these characteristics of plants and crops that anything that prevents individual plants and the crop as a whole from operating at their maximum photosynthetic efficiency will reduce yields and outputs. In theory, a typical crop could capture about 5 per cent of the sun's energy available to it; in practice the figure is closer to 1 per cent, partly as a result of climatic factors, such as cold or cloudy conditions, or through a shortage of water or nutrients, but mainly due to the effects of weeds, pests, and diseases.

A weed is proverbially a plant in the wrong place. Poppies (Papaver spp.) look pretty in the garden but are a problem in a wheat crop. The grass *Imperata cylindrica*, known as *lalang* in Malay, *cogon* in the US, and *blady grass* in Australia, is a traditional thatching material in South East Asia, and also cultivated for beach stabilization, and sometimes grazed in Africa, but is subject to official eradication schemes in several states of the south-eastern US. The significance of a weed can also depend on the time at which it grows: whereas an early weed infestation can smother a crop, late weed establishment in a fast-growing crop may have virtually no effect on yield. Thus the most obvious impact of weeds arises from their ability to compete for sunlight and so reduce crop photosynthesis, but they also compete with the crop for water, their seeds may contaminate the harvested crop, and they may make harvesting difficult, as for example when bindweed climbs up crop stems.

Almost any plant may therefore be a weed if it grows in the wrong place at the wrong time, but some plants are recognized as especially likely to cause problems. In temperate crops particular difficulties are caused by grass weeds, such as blackgrass and wild oats, annual broadleaved weeds, and perennial weeds. Fat Hen (*Chenopodium album*), for example, is now normally classified as an annual broadleaved weed, although in the past it has been used as a salad vegetable. The perennials include couch grass, bracken, docks, and thistles in temperate areas; in tropical areas one of the most troublesome weeds has been water hyacinth (*Eichhornia crassipes*), which was introduced from Brazil as an ornamental plant, escaped, and flourished to the extent that it blocked irrigation channels and caused floods.

Like weeds, pests are animals in the wrong place. The elephant that looks so magnificent on the plains of the Serengeti is a pest when it tramples through a crop; the rabbit may be welcome to nibble the grass in a woodland glade but not the young wheat on the other side of the hedge. At the opposite end of the size scale are minute eelworms (properly called nematodes) which can damage sugar beet and potatoes, and numerous insects, including aphids and locusts. They can attack all parts of the crop from the seeds to the roots. Some affect the plant's ability to photosynthesize by eating its leaves, and others its source of water and nutrients by eating the roots. Aphids have piercing mouthparts and feed on the sap in the plant cells, so stunting its growth, and in doing so some of them, such as the grain aphid (*Sitobion avenae*), pass on virus diseases (barley yellow dwarf virus or BYDV in this case) which further attack the plant.

Some of the most devastating pest attacks have happened when the causal organism has been introduced into a new environment. In the US the Colorado beetle (*Leptinotarsa decemlineata*) lived a quiet life on weeds of the nightshade family until about the middle of the 19th century. Then potatoes, which belong to the same botanical family as the nightshades (the *Solanaceae*) were introduced, and the population increased and spread. From Colorado, it reached the Atlantic coast of the US in 1873, crossed the Atlantic in 1901, and from there spread to continental Europe as a major pest of potatoes. Similarly the phylloxera aphid, to which American vines were relatively immune, caused havoc when it spread to continental European vineyards in the 19th century because the vine varieties used there, never having been exposed to it, had not evolved the same protective mechanisms as American varieties.

The range of insect pests is enormous, with almost every crop having several species that will live on it if environmental conditions are appropriate. African coffee crops, for example, are subject to the attacks of the variegated coffee bug, a capsid bug, the berry borer, mealy bugs, scale insects, stem borers, and leaf miners. Over the world as a whole, one estimate puts the loss of crops due to pests at as much as 25 per cent.

Diseases of crops are mostly caused by various species of fungi, but viruses and bacteria can also be responsible. Some fungi cause a loss of chlorophyll, making the leaf look yellow, and so reduce photosynthesis. Others grow in the vessels through which the plants transport water, making them wilt. An example of such a fungus which is currently causing considerable problems is *Fusarium oxysporum*, which lives in the soil and penetrates the roots of banana plants, eventually spreading so that the leaves turn yellow and the whole plant begins to wilt.

F.oxysporum was first identified in Panama at the end of the 19th century, and so is also known as 'Panama disease'. The variety that dominated the commercial banana trade at that time, 'Gros Michel', was susceptible to the disease, which by 1960 had spread to most producing countries. Fortunately, another variety, 'Cavendish', proved resistant to the races of *F.oxysporum* then existing, so it in turn became the dominant commercial variety, but in recent years a new race, termed Tropical Race 4 (TR4), has emerged from Malaysia and spread to other countries, and 'Cavendish' is susceptible to it. Potato blight, caused by the fungus *Phytophthora infestans*, is perhaps the best-known example of a plant disease that had an enormous socio-economic impact when it attacked the potato crop in Ireland in the middle of the 19th century.

How can we improve plant growth?

If plants are prevented from growing by shortages of light, heat, water, and nutrients, the obvious solution is to provide these things, and it is often possible to do that, but at a price. If the crop is valuable enough, the price may be worth paying. Some rich English landowners in the 18th century erected heated glasshouses in which to grow pineapples, and more recently police raids have found completely artificial growth rooms in which drug dealers have been growing cannabis plants.

In more common commercial circumstances tomato growers in northern Europe grow the crop in glasshouses, artificially heated, watered, fertilized, and in some cases with increased CO_2 levels, whereas in southern Spain the same crop needs no artificial heat but requires much more water to be provided from underground boreholes. Air travellers over dry parts of California can look down and see green circles in the desert that show where crops are being grown with the aid of large irrigation rigs circulating round a central water source (see Figure 4).

If transport costs can be kept low enough, it may be worth growing a crop where it is warm, for sale where it is too cold for it to grow: hence the Kenyan green bean in European supermarkets in winter months. Farmers everywhere have always tried to return organic nutrients to the land in the form of animal manure, and most farmers (except organic farmers) in developed countries, and increasing numbers in developing countries, now use artificial fertilizers (see earlier in this chapter).

Even in ideal growing conditions, crop growth may still be reduced, as we have seen, by the effects of weeds, pests, and diseases, and part of the skill set of the farmer is knowing how to reduce their impact. Weeds are most difficult to control, either by cultivation or

4. Centre pivot irrigation promotes crop growth in dry California.

herbicides, when they are very similar to the crop in which they are growing. If the weed looks like the crop one hesitates to attack it with a hoe, and if it resembles the crop physiologically and anatomically it will be difficult to find a herbicide that can kill it without also damaging the crop. Wild red rice (*Oryza rufipogon*), for example, is the most serious weed of rice, with losses in the Punjab at one time being estimated at over half of the crop.

Farmers have developed numerous strategies for weed control. It is important to sow uncontaminated seed at the right seed rate into a fertile seedbed to give the crop the best chance of competing with weeds; some weeds can be controlled by repeated cutting; and rotating crops enables broadleaved weeds that were difficult to control in a broadleaved crop to be tackled more effectively in a following cereal crop, and vice versa.

Before the introduction of chemical herbicides farmers and their workers would go through a crop with hoes, or pull out weeds by hand, and for high-value seed crops with fairly light weed infestations this is still sometimes done today. One of the reasons for inventing the seed drill, which planted seeds in rows, was that it was then easier to hoe between the rows with a horse-drawn hoe, as Jethro Tull argued in his book *Horse-Hoeing Husbandry* (1733), and crops with widely spaced rows, in which it is easier to employ inter-row cultivations, are often called 'cleaning crops'. But all these methods need a lot of labour.

At the beginning of the 20th century, before chemical herbicides were available, one farmer estimated that between a third and a half of the field labour on a farm was devoted to the destruction of weeds. It is therefore easy to see why farmers, especially in high-wage economies, have been such enthusiastic adopters of herbicides, and why organic farmers, who cannot use them, often argue that weeds are more difficult to manage than pests and diseases.

As with weeds, so with pests and diseases: farmers try to use a mixture of management and chemical methods to reduce their impact. Rotating crops, so that different ones grow on a field in succeeding years, helps to break the life cycle of fungi and soil-borne insects, and sowing early or late to miss the time at which the pest or disease is most active, are examples of management methods, and farmers increasingly resort to expensive agrochemicals only when these methods are ineffective.

Chemical control is itself not without problems. Not only may the target organism develop resistance to it, but it may also have unwanted side effects on non-target species. The impact of extensive use of chlorinated hydrocarbon insecticides on the breeding success of birds of prey is one well-known example. One solution to this problem is to use predators to attack pest species, as in the use of parasitic wasps to attack glasshouse whitefly; another is to attempt to breed resistance to attack into the plant.

Pest and disease resistance is just one of the possible objectives of the plant breeder. Essentially, most breeding programmes attempt to increase crop yield, or quality, or both, although sometimes increased yield is obtained at the expense of quality: for example, IR8, one of the early high-yielding rice varieties, released in 1966 by the International Rice Research Institute in the Philippines, was disliked by consumers because the grain was chalky and brittle.

There are various ways in which breeders can attempt to increase yield. They can produce varieties that are resistant to drought or to common pests and diseases, such as diseases of cereal foliage. They can also try to increase the saleable part of the plant at the expense of the rest. This has been extensively used by both rice and wheat breeders, who have produced varieties with shorter straw but bigger ears. The total biomass may not change much, but more of it can be sold.

Other objectives depend on the particular characteristics of individual crops. Older sugar beet varieties produced seed in clusters, which would produce three or four plants when they germinated. These had to be laboriously singled by hand hoe to produce one strong plant, so when breeders produced 'monogerm' seed, with only one seed per cluster, the need for this job was removed and costs were correspondingly reduced. Some varieties of the other source of sugar, sugar cane, have sharp leaf edges, which can be painful for workers in the cane fields, so breeders have attempted to produce varieties which minimize this characteristic. Modern plant breeding methods and their implications are discussed in Chapter 6.

Crops and cropping

'Most of us have a potato-shaped space inside that must be filled at every meal, if not by potatoes then by something equally filling' wrote Katharine Whitehorn in her classic *Cooking in a Bedsitter* (1963). These staple foods are usually starch-based, and come from some of the most widely grown crops, such as rice, wheat, maize, potatoes, cassava, and yams. The science fiction writer John Christopher, in his novel *The Death of Grass* (1956), suggested that world civilization would break down if a disease began to kill off all the *Poaceae*, the family of plants that includes not only the grasses, which are the most widely grown crop in the world, but also the temperate cereals (wheat, barley, oats, rye), the tropical and semi-tropical cereals (maize, rice, the millets, sorghum), and sugar cane. We should also be short of proteins if a similar plague affected the legumes (*Fabaceae*), the family that includes the peas, beans, and lentils, and of vegetables and oilseeds if the *Brassicaceae* (cabbages, rape, swedes, mustards) were attacked.

Some crops produce more than one product. Wheat makes bread and barley is malted for beer, but both are also used to feed animals, especially pigs and chickens. Soybeans can be crushed to produce oil, and what remains is then a high-protein animal feed.

Linseed stems are the source of flax fibre to make linen, while the seeds are the source of an industrial oil, although over time different varieties of the species have been developed to specialize in one or the other product.

There are different varieties of most crops, hundreds of different varieties in the case of wheat, for example. Some have been bred to grow better in hotter and drier climates, some to produce better breadmaking flour, others to be resistant to various fungal diseases, and so on. The varieties of maize that are grown in tropical and semi-tropical countries and form one of the most important sources of dietary starch would not be those grown to produce maize silage for animal feed by farmers in north-west Europe, or to produce sweetcorn by American farmers. The grape varieties that produce the great wines are not designed to be eaten as fresh table grapes.

These are all human or animal food crops, but there are also extensive areas of crops such as cotton, jute, and hemp which are grown for their fibres, and smaller areas of borage, calendula, and evening primrose for industrial and pharmaceutical oils. Increasingly, crops may be grown to produce biomass, or fuel such as bioethanol.

All these species may also be classified according to their mode of growth. There are tree crops, such as coffee, cocoa, bananas, olives, and apples which, once established, grow for many years. Other perennials may be bush crops, such as grape vines, currants, or raspberries, which again can remain in place, for many years in the case of vines. But the majority of crops that farmers grow are annuals or (a few) biennials. In temperate climates that means the crop usually grows during the summer and is harvested in the autumn.

The farmer prepares a seedbed, sometimes by ploughing to bury the remains of the previous crop, then cultivating to break down the clods of soil, before depositing the seeds in rows using a seed drill. Some cereals such as wheat or barley may be sown in the autumn and become established before the really cold weather occurs. They are then ready to grow when the soil warms up in the spring, and so maximize the period when they can photosynthesize most effectively. The time available to prepare a seedbed in the autumn may be limited, or winters so harsh that many seed crop varieties are sown in the spring. In northern Canada, for instance, crops may have only ninety days in which to complete the cycle from sowing to harvest. Root crops such as potatoes and sugar beet are normally planted in spring for harvest in the following autumn. Once the crop is established it may be given one or more dressings of fertilizer, and herbicide and fungicide sprays. Harvesting takes place when the grain has filled and matured, in the case of cereals and oilseeds, and when little further growth is expected in the case of potatoes and sugar beet.

This is obviously a very basic outline of the process, and it should be clear that it involves many different decisions, about the best way to prepare the seedbed, which varieties to plant, what seed rate to use to produce the optimum plant population, how much fertilizer to use and when to apply it, which weeds, pests, and diseases might threaten the crop and how to combat them, when to harvest, and so on, all of which may vary from farm to farm, field to field, and even within the same field.

Farmers also have to decide which combination of crops to grow. Sometimes they grow one crop year after year, but this can lead to a build-up of pests and diseases, which have to be controlled, and a depletion of soil nutrients, which have to be replaced. There is a balance to be struck between growing the most profitable crop, and maintaining crop health and soil fertility.

A modern six-course arable rotation in Europe might see two years of wheat followed by one of beans, followed by wheat, then barley, then oilseed rape (canola). The beans break the wheat pest and disease life cycles and, being legumes, help to fix nitrogen in the soil; the oilseed rape also breaks the cereal disease life cycle. The classic Norfolk four-course rotation had wheat followed by turnips, which were heavily dressed with farmyard manure and also carefully weeded, so that there was a rich and weed-free soil for the following barley crop, which was then followed by clover, which fixed nitrogen and so enriched the soil for the following wheat crop, and so on. Modern fertilizers and pesticides have enabled farmers to increase the time between the restorative crops, but the principle remains applicable, and cheaper than complete reliance on chemicals. Rotations are simplified rather than eliminated.

In the Midwestern states of the US, a region often called the Corn Belt, farmers now rely on a rotation of corn (confusingly called maize in Britain, where 'corn' is a collective noun for wheat, barley, and oats) and soya beans. In the southern states, corn/maize may be grown in rotation with cotton and a legume such as cowpeas. A variation on this theme may be found in some tropical countries, where a crop such as cassava, which stays in the ground for several years, may either be grown in a pure stand or intercropped—grown in association with—annual crops. In some rice-growing areas the land may remain fallow until the next rice crop is sown, but in some parts of South East Asia, where the soil is suitable for cultivation in the dry season, the rice may be rotated with pulses, maize, or vegetables.

When farmers decide on the best crops to put in a rotation, they also have to think about the ways in which they can harvest them: all the crops in the wheat/beans/barley/rape rotation mentioned earlier can be combine harvested, whereas inserting a grass break would require the farm to have grazing animals, or additional machinery to cut the grass and make it into hay or silage.

The decisions that farmers make about the crops they grow therefore depend upon a wide range of factors: not only what the soil and climate will allow them to grow, but also what they have the knowledge and equipment to produce, and what they or their animals need to eat or can sell. These considerations are discussed in the following chapters.

Chapter 2
Farm animals

Humans have domesticated a relatively small proportion of the available animal species. As the American polymath Jared Diamond has argued, animals needed particular characteristics to be suitable for domestication: the right diet, usually based on a wide range of plants, a relatively rapid growth rate, an ability to breed in captivity, and a suitable disposition and social organization, so that they don't panic and have a well-developed dominance hierarchy in which humans can take over as leaders. Only a few species combined all of these characteristics, and they have become the most common farm animals. The top ten (i.e. the most commonly kept) farm animals are shown in Table 2.

These numbers need to be used with care, because obviously one chicken produces nothing like the same amount of food as a cow, a sheep, or any of the bigger animals. They also need to be set in a time context. The number of chickens and ducks in the world roughly doubled in the twenty years since 1992, and goat numbers increased by 65 per cent, whereas cattle and pig numbers only went up by about 13 per cent, and sheep numbers were static. Furthermore, farmers keep other animals that do not appear in this list, such as the llamas, alpacas, and guinea pigs of South America, yaks, and even insects such as silk moths and honeybees.

Table 2 World farm animal population 2012 (million head)

Chickens	21,867	Pigs	966
Cattle	1,485	Turkeys	476
Ducks	1,316	Geese and guineafowl	382
Sheep	1,169	Buffalo	199
Goats	996	Camels	27

FAO.FAOSTAT. Live Animals: Number of Heads (FAO 2015) Accessed 29 July 2014 <http://faostat3.fao.org/faostat-gateway/go/to/browse/Q/QA/E>.

Some animals, such as reindeer, muskoxen, or ostriches, are best described as semi-domesticated, as they are exploited by people but are not managed as intensively as most farm livestock. Others, such as horses and donkeys, are used in farming, but also in non-agricultural transport roles, as are cattle in some parts of the world, so that it is difficult, and not always useful, to measure the numbers of livestock kept for purely agricultural purposes.

Many farm animals have multiple functions. Cattle produce milk (and blood in some parts of Africa) and in some parts of the world pull ploughs and carts while they are alive, and produce meat and leather when they are dead; sheep are shorn for their wool and produce milk as well as being slaughtered for meat, and camels and buffaloes produce milk and pull ploughs.

It is perfectly possible to live on a vegan diet, with no animal products at all, and in developed countries we no longer rely on animal fibres for clothing or animal power for traction, so why do we bother to keep animals at all? Is it simply that animals produce a convenient collection of nutrients that are not always easily available to vegans? In part at least, it must be a question of tradition. Before the invention of steam and internal combustion engines, many animals—horses, donkeys, mules, cattle, and even dogs—were kept for their muscle power, to pull carts, ploughs, and other agricultural implements. Their milk and meat were a by-product for which people developed a taste. Pigs and chickens, to which this argument clearly does not apply, were valued as scavengers that would provide meat and eggs by eating food waste or things like acorns or insects that were unpalatable to most humans. Thus, over the course of thousands of years of living with domesticated animals, humans have developed a taste (and are willing to pay) for the meat, eggs, and dairy products that they produce, and have found innumerable ways of combining them together and with plant products to make the food we eat. Farmers are therefore interested in making their animals grow quickly, and put on the right combination of muscle and fat that butchers prefer, or produce lots of milk or eggs or wool, and this chapter is about what they need to know in order to do that.

Feeding farm animals

Some basic principles hold true for all kinds of animal husbandry (as the care of farm animals is often called). Boiled down to its essentials, and forgetting for a moment that farm animals are sentient creatures, with all that implies, animal husbandry is about converting the carbohydrates, proteins, fats, etc. in plants into the carbohydrates, proteins, fats, etc. in animals, so that they can give us meat and milk, and, in some cases, have the energy to pull our farm implements.

Animals possess the enzymes (biochemical catalysts) needed to break down the starch in plants into its component glucose molecules. Glucose is used directly to provide energy for the animal, or synthesized into glycogen for energy storage. In the mammary gland, it is combined with other molecules to form lactose, the main carbohydrate in milk. Mammals and birds do not, however, possess the enzymes necessary to break down the cellulose (dietary fibre) in plant tissues into glucose, although some bacteria and fungi do. This means that those animals that have evolved to live with a gut flora containing the right sort of bacteria can extract nutrients from relatively fibrous leaves, containing a high proportion of cellulose. Those that have not must eat seeds and tubers containing the starch that they can metabolize.

This is a vital difference, because it means that the cellulose metabolizers (such as cows, sheep, goats, camels, and buffalo) do not have to compete with the starch metabolizers (such as chickens, pigs, and people) for food. In practice, of course, the carbohydrates in the diets of both groups include a mixture of sugars, starch, and cellulose, but what happens to them depends upon the gut anatomy of the various animals.

An animal's muscles, skin, hair, or wool, and even part of its bones, are made of proteins, which are large molecules formed by linking simpler compounds, called amino acids, of which there are about twenty different examples. Plants and bacteria can synthesize all twenty, but there are ten of them that cannot be synthesized by animals. These are known as essential amino acids, and they must be provided in the diet. Put simply, the digestive metabolism of the animal involves breaking down the proteins in its food into their constituent amino acids, which are then available to form the different proteins in the animal itself.

The other major constituents of the animal's body are the various fats, which are not only deposited under the skin, but are also found within and between muscles and muscle fibres, and also form parts of nerves, cell membranes, and hormones. Again, most of these can be synthesized by the animal from the fats and oils in its food. There are also minor nutrients, in terms of weight, which are nonetheless vital components of an animal's food. These are the minerals and vitamins, the lack of which prevents efficient metabolism and is usually demonstrated as disease, such as bone deformations (rickets) caused by vitamin D deficiency.

What happens to food after it has been eaten depends upon the anatomy of the animal's digestive tract. Simple-stomached (monogastric) animals begin the process of chemical breakdown of food in the stomach, after which the semi-digested material passes into the small intestine, where most of the remaining digestion of carbohydrates, proteins, and lipids occurs. What remains is then moved by muscular contractions into the large intestine, where water is absorbed before the undigested waste is voided in the process of defecation. This is in essence what happens in humans, who, as mentioned earlier, have little ability to digest cellulose.

The digestive tract of the pig is similar, except that the caecum (the part of the gut at the beginning of the large intestine) is larger than in humans and has a microbial population that can break down cellulose to a limited extent, so that up to 20 per cent of the pig's energy may be derived from this source. This means that pigs can derive some food value from grass and other vegetable fibres that would pass straight through a human.

The digestive tract of the chicken is completely different in detail from that of monogastric mammals up to the small intestine, but from a farmer's viewpoint the result is much the same: it needs a relatively low-fibre diet. In horses and donkeys, however, the caecum is much enlarged, and its microbial population has a considerable ability to break down cellulose. Roughages such as grass and hay can therefore form a large part of equine diets.

Ruminants, such as cattle, sheep, goats, and buffalo (and, with a different anatomy, camels), also carry bacteria that can break down cellulose, but in an enlarged and complex stomach called the rumen. After eating, ruminants can bring food back into their mouths and chew it again—the process of rumination, or chewing the cud—which helps to break down fibrous food for further metabolism by the rumen bacteria. In fact, without some dietary fibre, ruminant digestion does not work properly, which means that although highly productive cattle, for example, may be given concentrated foods containing starch, proteins, and fats, they also need fibrous material if they are not to suffer digestive problems. Furthermore, in milking cows high-concentrate/low-forage diets can depress the fat level in the milk.

These differences between ruminants and monogastrics are important from a farmer's viewpoint, because they mean that ruminants can live and thrive on a high-fibre diet of plant leaves on which a monogastric animal would struggle to survive. Consequently, they can utilize grass, which makes a good break crop on arable land (see Chapter 1), and can also exploit the vegetation on land that is too steep, rocky, or poor to be cultivated.

The way in which a farmer feeds animals therefore depends on their species and consequent anatomy and physiology, the feed available, and the purpose for which they are being kept. All animals have a 'maintenance' requirement for the nutrients necessary to maintain their body weight and enable their various physiological functions, without which they will eventually starve and die. Most of them also have a 'production' requirement, for growth, pregnancy, lactation, and activity.

In animals that are to be slaughtered for meat, feeding for growth is usually the major cost of production. In the young animal it is the bones that reach their peak growth rate first. As the animal ages, and the relative growth rate of bone decreases, it is the muscle growth rate that peaks, and finally, as muscle growth begins to slow down, fat deposition reaches a maximum (see Figure 5). Higher levels of fat are generally found in older or more rapidly growing animals within a breed.

Farmers try to sell their animals for slaughter when their fat levels are at an optimum: too little fat, and their meat is likely to be dry and flavourless; too much, and the animal loses value. But what constitutes too much can change over time. In the 19th century,

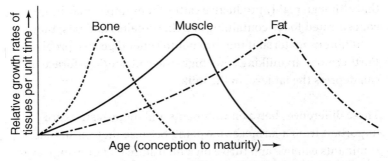

5. **Relative growth rates of an animal's body tissues from conception to maturity.**

for example, pigs were often fed to much greater weights than they would be today, because there was then a bigger market for their lard, or subcutaneous fat.

When an animal becomes pregnant, her production requirement obviously increases because in addition to maintaining her own body she also has to cope with the requirements of her growing offspring. Although this ceases after its birth, she then has to produce milk to feed it, which again needs additional nutrients over and above her maintenance requirements. In practice, both pregnancy and lactation may occur at the same time in dairy cows, because they become capable of conceiving again two or three months after giving birth, whereas their lactation can last for nine or ten months. In a high-yielding dairy herd the farmer will try to get the cows to have a calf every year, so that as the milk yield declines from its peak, the requirements of the growing embryo increase. High-yielding cows may need three times their maintenance requirement or more for these production requirements.

It is difficult to get enough nutrients into cows that are producing a lot of milk using forage alone, so they are usually given an additional ration of high-energy and protein concentrates. Even so, many high-producing cows will still not receive sufficient nutrients for their needs and will utilize backfat reserves. On the other hand, ewes on fresh spring grass, readily digestible and with a high protein content, may be able to feed their lambs without much if any additional concentrate feeding.

The corresponding situation for chickens is egg production. Although most hens could probably find their maintenance ration by foraging around a farmyard, a hen in a high-yield production system laying an egg almost every day needs a diet with a high nutrient concentration. Similarly, a horse needs a feed of oats or oats and beans if it is to pull a plough all day, whereas it needs little more than grass if it is not working.

Breeding farm animals

In addition to feeding their animals for maximum performance, farmers can also affect their output by the way in which they breed them. It is important to distinguish between a breed and a species. Essentially, species cannot interbreed: the joke about producing woolly jumpers by crossing a sheep and a kangaroo is a biological impossibility. But within the same species animals of very different appearance can be successfully crossed. Fertilizing a female horse (mare) by a male donkey produces the mule which has been used as a pack and traction animal for centuries, and Indian humped zebu cattle (*Bos indicus*) can be successfully crossed with European-type cattle (*Bos taurus*). In the US, for example, at the beginning of the 20th century, breeders crossed Brahmans, a zebu breed, with Angus cattle originating in Scotland, to produce the Brangus beef breed. There have even been crosses of *Bos taurus* with the North American buffalo or bison (*Bos bison*) to form a hybrid known as the cattalo or beefalo.

Within a species, therefore, one breed is distinguished from another by differences in colour, size, and/or shape which are maintained from generation to generation. These differences probably originated in geographical variations, when farmers in one region produced animals which suited their particular local resources and needs. As communications improved, farmers elsewhere realized that they could improve their own stock by using those from another area. In the 18th century, for example, farmers in the north-east of England imported bulls from the Netherlands which 'did much service' in improving local cattle. An early 20th-century survey of British livestock listed thirteen major breeds of cattle, eight of pigs, and twenty-nine of sheep, in addition to various crosses and minor breeds. From the 1960s onwards further breeds of cattle, such as the Holstein, Charolais, Limousin, Simmental, and several others were imported from continental Europe.

Different breeds are bred for different purposes. Specialist dairy breeds, such as the Holstein, are not especially good at producing meat, and cows of a beef breed such as the Aberdeen Angus have not been bred to produce large quantities of milk. But these specialist beef animals will begin to lay down subcutaneous and intramuscular fat at lower weights than the specialist dairy breeds, so that they can be 'fattened off grass'. In other words, they reach slaughter weight without a lot of concentrate feeding.

Merino sheep have been bred to produce large quantities of fine wool, but give little milk, few lambs, and poor meat, whereas a prolific Mule or Scottish Halfbred ewe, crossed with a Suffolk or Texel ram, has enough milk to raise twin lambs of high carcass quality. Sheep breeders in the US distinguish between ewe breeds such as the Rambouillet, Corriedale, Targhee, or Border Leicester, noted for their prolificacy, size, milking ability, and longevity, and ram breeds such as the Suffolk, Cheviot, or Texel, which possess good growth rate and carcass characteristics.

Modern hens bred to lay large quantities of eggs do not appear on supermarket shelves as oven-ready chickens at the end of their laying lives. They are 'spent hens', useful only for processing or pet food. Riding horses and ponies, and riding camels, are of a different shape and size from carthorses and pack camels. All of these thousands of different domesticated breeds are descended from the same original wild ancestors, but over thousands of years of selection by humans an enormous range of animals with different shapes, sizes, and attributes has emerged.

Traditional breeders worked by eye and memory. That is to say, they chose the males and females that they wanted to mate according to the characteristics they wished to produce. They expected a larger than normal bull mated to a larger than normal cow to produce big fast-growing offspring, or a high-yielding dairy cow mated to the son of another high-yielding cow to produce high-yielding daughters. Their expectations were fulfilled often enough for them to produce the traditional breeds that are still to be seen on modern farms.

By the end of the 19th century in Britain there were breed societies that kept records of the ancestry of individual animals, and shows in which breeders competed against each other to produce the best horses, cattle, sheep, and pigs, which sold for high prices. The ideal animal for the breed was described in terms of its colour and shape. In the 20th century this process was taken further by recording milk yields, which enabled breeders to select animals to mate on the basis of the performance that they wished to maximize. This was a form of performance testing, which was obviously more difficult to carry out on meat-producing animals. Carcasses don't breed! However, it is possible to weigh both animals and their food, and so measure their growth rate and feed conversion efficiency, and within the last forty years or so ultrasonic scanners have been developed which can measure the backfat level on live animals, so enabling assessment of their carcass composition.

Measurement of characteristics which are not expressed in the animal, such as the effect on milk yields of the genes transmitted by male animals, can be measured by testing progeny. For example, a bull's genetic ability to sire high-yielding daughters can be assessed by comparing the yields of his daughters with those of other bulls in the same herds, a process known as contemporary comparison.

Some of the characteristics of an animal are controlled by a single pair of genes. The black coat colour of an Aberdeen Angus, for example, is always dominant over the red coat of a Hereford, so Angus/Hereford cross cattle are always black with a white face. Most of the economically important attributes of animals, however, are controlled by many genes, so the outcome of a single breeding event is less predictable.

Three factors therefore have a big influence on the extent to which breeders can bring about genetic change in an animal population: heritability, gestation period, and the number of animals available. Heritability expresses the proportion of any difference between the individual and the population average that can be attributed to genetics rather than environment, on a scale of 0 to 1. A high score means that the offspring resembles the parent. For example, the heritability of the lactation yield is between 0.2 and 0.3, whereas that of the fat content of the milk is between 0.5 and 0.6. As a result it is easier to change the average fat content in the milk of a dairy herd than it is to increase its output.

The gestation period—the length of time the offspring spends in the womb (see Table 3)—also matters, because the faster a new generation is available to breed the more rapidly the breeder can bring about genetic change. Finally, the greater the number of animals available to the breeder, the greater the selection pressure can be. The best animal selected from one hundred is likely to be better than one selected from ten. We would therefore expect genetic improvement to be more rapid in poultry and pigs than in cattle. This is because two generations of pigs or poultry can be produced in a year, and many animals can be kept on a small area.

Table 3 Approximate gestation periods

	Average days	
Camel	406	12–15 months
Horses	340	more than 11 months
Buffalo	300–40	10–11 months
Cattle	283	9 months 10 days
Sheep	150	about 5 months
Goats	156	about 5 months
Pigs	116	3 months, 3 weeks, 3 days

It can be as long as three years between one cattle generation and another, and only a couple of animals can be kept on each hectare.

This rapid change in pigs and poultry is indeed found in practice. In the US, between 1947 and 1988, poultry breeders halved the time and the quantity of feed needed to get a broiler chicken ready for slaughter. A 2 kg broiler can now be produced in six weeks. Over roughly the same period the growth rate of beef animals increased too, but only by about 16 per cent. Most commercial pigs and poultry are now bred by specialist breeding companies employing specialized geneticists.

Cattle breeding has not yet gone as far in specialization, but artificial insemination (AI) has helped dairy farmers, especially, to have access to high-quality genetic material. It means that rather than having to keep their own bull, they can use semen (which arrives in frozen form) from a bull with tested breeding performance. Since AI uses bulls selected under high selection pressure, genetic improvement is more rapid than is achieved using natural service. Additionally, AI allows superior males to be used on many more females. Some specialist breeders also use embryo transfer techniques, in which multiple embryos from high-quality cows are flushed out of the original cow and transplanted individually into recipient animals, thus allowing lower-performing cows to have high-quality offspring.

Housing farm animals

In addition to feeding and breeding their animals, farmers also attempt to increase their productivity by housing them. Farm animals are quite capable of living outside all of their lives, although hens like to lay their eggs in nests, and some sows living outside will, if they can find enough suitable material, build large nests in which to give birth to their piglets. But having fed their animals, farmers are often reluctant to see part of the energy content of the ration going into keeping them warm during periods of cold, wind, and rain. This is a particularly deadly combination for young lambs born outside, even though they are perfectly capable of surviving on a clear, still, cold night. Conversely, in hot climates animals may need protection from the sun, and water buffalo need access to water in which to wallow during the heat of the day.

The sort of accommodation provided on farms is enormously variable. At one extreme there are simple buildings, little more than a roof, to keep the rain off; at the opposite extreme are controlled environment houses with artificial lighting, ventilation, and temperature control in which fattening pigs and broiler chickens may spend the whole of their lives.

In a sense this level of housing is as much for the benefit of the farmer as the animals. In an egg battery house it is easier to automate the collection of the eggs and the feeding of the birds than it is for a flock of free-range hens, thus reducing labour costs. In addition, egg laying by hens is stimulated by increasing day length, so if the birds are reliant on artificial lighting the farmer has greater control over their laying pattern. Even simple accommodation for cattle means that the farmer can decide where the farmyard manure that they produce will be used, and indeed this was one of the early arguments for housing fattening cattle. However, there is little doubt that housing, especially intensive housing for the whole of an animal's life, affects its behaviour, and this has led to arguments about animal welfare which are discussed further in Chapter 5.

Housing also has health implications. Although a farmer may house animals to protect them from the weather, the result may be to expose them to increased risks of disease. It is common, for example, to treat housed poultry with drugs to inhibit the development of various species of the *Eimeria* parasite that cause coccidiosis, an infection of the intestine. Crowded housing may also induce stress in animals, as does hunger and thirst, and a stressed animal is more vulnerable to disease.

Farm animal health and disease

Farm animals can suffer from as wide a range of diseases as humans, with the added difficulty that they cannot tell the farmer when they are feeling unwell. One of the attributes of a good livestock manager is the ability to spot the early signs of disease, so that the animal can be isolated and/or treated, and to know the circumstances that might predispose the animal to problems.

Well-fed, high-yielding dairy cows sometimes collapse with little warning a few days after calving, a condition known as 'milk fever', which, left untreated, can be fatal. The cause is a fall in the level of blood calcium, which is easily treated, as long as the farmer knows what to do, by an injection of calcium borogluconate. This is just one example of the problems that shortages or surpluses of various minerals or vitamins can cause.

Animals are also subject to the attacks of ectoparasites (external parasites) such as flies, lice, and ticks. In the south-eastern states of the US, for example, the maggots of the screw-worm fly

(*Cochliomyia hominivorax*) burrow into the flesh of cattle. There was a successful control programme for the fly which involved releasing laboratory-bred sterilized male flies at the peak of the mating season, so that the population was reduced in a few generations and by the 1980s it had been eradicated.

Attempts to control tsetse flies (various species of the genus *Glossina*) in Africa have been less successful, and there remain large areas where cattle cannot graze for parts or all of the year. The problem is not so much the fly itself, as its ability to infect those it bites with trypanosomes, small flagellate organisms that live in the blood and cause sleeping sickness in humans and a related condition (one name for which is *nagana*) involving anaemia, fever, and slow progressive emaciation in cattle.

Similarly, ticks, of which there are numerous species, may be carriers of various diseases, such as redwater fever, East Coast fever, biliary fever, and gall sickness in African cattle, which therefore need to be sprayed weekly to control the ticks. There are also internal parasites, such as liver fluke and roundworms, bacteria, such as the clostridial bacteria that cause tetanus, and viruses, the causes of cattle plague and foot and mouth disease. All of these are potential sources of farming problems.

In the last half century or so the veterinary profession and the pharmaceutical industry have created numerous products to help farmers to combat these diseases. There are anthelmintics to control roundworms, vaccines against clostridia, and antibiotics against all kinds of bacterial infections. The question that individual farmers have to answer is whether the cost of treatment is more or less than the loss of profit from a disease, and there seems little doubt that in some circumstances they take the decision that it is worth accepting the risk of some deaths in their flocks and herds rather than bearing the cost of preventative medicines.

For some diseases, however, such as foot and mouth disease, the state often takes the view that the potential effects of disease are so great that something must be done about it, up to and including the slaughter of the affected animals. This is especially so in the case of 'zoonoses', diseases that can potentially spread from animals to people. In the last few years there have been numerous examples of these in Britain: bovine spongiform encephalopathy (BSE), Highly pathogenic avian influenza (HPAI, or bird flu), *E.coli* O157, *salmonella*, *campylobacter*, *listeria*, and bovine tuberculosis (bTB).

Animal production systems

Taking all these considerations of feeding, breeding, housing, and disease control into account, the farmer then has to decide upon a system of production. There are almost as many different ways of keeping animals as there are different farmers, but most of them can be classified in terms of intensity, specialization, and scale. An intensive system is one which concentrates animals on a relatively small land area but uses a lot of labour, equipment, housing, and concentrated feedstuffs, whereas an extensive system uses more land and fewer of the other inputs. Thus most commercial pig and poultry units in developed countries would be described as intensive, whereas nomadic cattle herders in east Africa, or cattle ranchers in the Dakotas in the US, operate extensive systems.

Some specialized livestock farms may have only pigs or poultry, whereas on mixed farms there may be several animal and crop enterprises, as in the traditional mixed European farm, with dairy and beef cattle, sheep, a few pigs, and poultry, all fed from the grass, root crops, and cereals produced on the farm. A specialized Western European dairy farm, in contrast, would only have dairy cows, perhaps with some young stock as herd replacements, with most of the land devoted to grass, and some maize for silage.

The scale on which livestock farming takes place is almost infinitely variable, from the part-time holding with a few pigs and chickens to large units with thousands of animals on thousands of hectares. Nevertheless, most livestock farmers, no matter what the scale, intensity, or degree of specialization of their holdings, will have some of the same jobs to do.

All livestock farmers need to ensure that their animals have access to food and water every day. For some this is easy, because their animals are grazing on fields or range lands, and they may not even check them every day. Or it may simply involve pushing a button to bring automatic feeders into operation. But for others, especially in winter in temperate climates, or in the tropical dry seasons, it may take up much of the day, needing them to move large quantities of hay or silage and concentrate feeds.

If they have milking animals, they also need to ensure that they are milked every day. A cow gives milk for nine or ten months after she has had a calf (for a goat it can be much longer, and some will even give milk without having had young), and the simplest way of removing the milk from the udder is to leave it to the calf to do the job. This is the way some beef producers, with what are called suckler cows, operate, and it gives the calf a good start in life, but obviously it produces no milk for human consumption or sale. Moreover, the calf needs nothing like the milk output of a modern West European specialist dairy cow, averaging over 7,000 litres in a 305 day lactation.

For thousands of years, therefore, farmers have been restricting the calf's access to the cow, and drawing off at least some of her milk for themselves. For most of that time they have done it by hand, sitting by the side of the cow in the morning and late in the afternoon and using the actions of their hands to imitate the sucking of the calf. It was not until the 20th century that a successful milking machine was invented, but once it was available it allowed farmers to keep a larger number of cows without needing to employ many more milkers. In recent years completely automatic milking machines have been developed, which means that the cows can choose for themselves when they wish to be milked. Many of them choose to be milked more than twice a day.

Livestock farmers need to be aware of the normal behaviour of their animals. Cows reveal their readiness to be mated by their willingness to be mounted by other cows in the herd. If they are running with a bull this is mostly of interest to him, but it matters to the farmer when the cows are artificially inseminated, because getting the insemination at the right time, when the cow is 'on heat', determines whether or not she will conceive. Heat detection in sheep is much more difficult for humans, and has to be left to the ram.

Knowing when an animal is about to give birth is also important. Cows will often go away from the herd into a corner of a field, and sows kept outdoors will collect bits of straw and twigs as if they were going to build a nest. In large herds this individual observation is not always possible, and pregnant animals may be moved into calving ('farrowing' is the equivalent English term for pregnant sows) accommodation according to the length of time since they were mated. Abnormal behaviour can also be a sign that an animal is suffering from disease, and part of the farmer's job is to carry out routine injections to vaccinate animals or kill their intestinal worms, or to inspect, and when necessary trim and medicate, sheeps' feet to prevent lameness.

Cleanliness is an important part of disease control, and farmers often spend a lot of their time moving dung or cleaning housing between batches of animals so that infective organisms are not transferred from one lot to another. They will also move stock to clean pasture for the same reason, and ensure that they stay in the correct fields, which means time must be spent on the maintenance of fences and hedges.

A farmer also needs to know when animals destined for the meat market are ready for slaughter, which might simply be a matter of checking that they have reached the weight that the slaughterhouse has specified, but can also require some assessment of the subcutaneous fat cover by feeling the bones through the skin along the back.

When all those tasks have been completed, grazing livestock farmers have to ensure that they have conserved enough fodder during the growing season to last their animals though the winter or the dry season.

Although many modern animals, especially pigs and chickens, but increasingly also dairy cows, may never see a green field, their basic biology still determines the way in which farmers use them to produce food and fibre. As such, they can never forget that they are living organisms, and not simply commodities.

Chapter 3
Agricultural products and trade

As we have seen in Chapters 1 and 2, farmers spend much of their time growing crops and raising livestock, usually in excess of their family's needs. Part of their time has to be spent selling the surplus. This chapter is about what the agricultural industry as a whole produces, and about the working of the markets into which its output is sold, both local and international.

It's obvious that farmers produce food for themselves and other people, but they also produce livestock feed for farm animals (and companion animals—pets—too). These are the two biggest categories of farm product, but the list goes on: plant fibres, such as cotton or jute; animal fibres, mostly wool; fuel in various forms, from biogas to dried camel dung; pharmaceuticals, such as insulin extracted from the pancreas of pigs, and Cerebrolysin from their brains, used in the treatment of Alzheimer's disease. Where the list stops depends upon where we put the boundary around what we call farming. It would certainly include tobacco growing, but what about the cultivation of opium poppies? We perhaps think of flowers as being grown by gardeners, or, on a commercial scale, on smallholdings by people we describe as 'market gardeners', but increasingly they are grown on a field scale, as anyone who has driven round parts of the Netherlands will have seen; one might almost say on a factory scale, given the size of some of the greenhouses there. The same applies to many kinds of salad crops, such as tomatoes, peppers, and lettuces. And what about plantation crops, such as pineapples, tea, or rubber trees? Or those from small trees or shrubs, such as coffee or cocoa?

It is also important to remember that farmers don't just produce physical products. They are responsible for providing what are often classified together as 'ecosystem services': farmland is a habitat for many different kinds of wildlife, farmers may use (or allow other people to use) their land as a site for solar panels or wind-powered electricity generation, and many farming landscapes are beautiful to look at, so that farming produces a resource for the tourism industry.

The relative importance of fuel, pharmaceuticals, ecosystem services, and so on, as compared to food production, is difficult to determine, certainly on a world scale. But there is little doubt that food is the main and most important output of farming. The smaller the farm, and the less developed the agricultural economy, the more likely this is to be the case. Vegetables, salad crops, fruit, nuts, potatoes and other roots, milk, eggs, and chickens can all be grown and processed to the stage at which they are ready to be eaten without needing any more equipment or expertise beyond that normally available within the farm household.

However, as Table 4 shows, the great bulk of the world's agricultural output comes in the form of products that need more processing than most households can now manage. Moreover, since there are now more people living in towns and cities than in the countryside, the majority of people will not be able to grow their own food. Therefore, as soon as farmers produce more food than their immediate family can consume, they should be seen as suppliers of raw materials to the food industry, and the greater the scale on which they operate, the more likely this is to be the case. Even potatoes and onions, which are ready for the kitchen as soon as they leave the ground, or apples, eatable fresh from the tree, need to be stored to increase the season for which they are available, and transported to a place where the consumer can buy them.

The more industrialized the economy, the more complex that apparently simple process becomes. Apples will be sorted, graded, and perhaps have those little labels that tell you their variety stuck on them. They may then need to be stored, perhaps for several months, before being sent to a central distribution depot, and then finally to an individual shop or supermarket.

At the opposite extreme, entire industries have developed to process cereals, milk, and meat. Farmers sell their wheat to millers, who grind it into flour (and various grades of bran that can be sieved out and sold as animal feed) which they then sell to bakers. If the wheat is not of the right variety or quality to be milled for bread or biscuit flour, it will be sold to animal feed manufacturers. Milk may be consumed straight from the cow or goat, but for people in developed economies it will have been chilled on the farm, transported from the farm to the dairy, pasteurized or sterilized, and put into bottles or plastic containers. Not all of it will be needed for the liquid market, so it will be converted into butter, or yoghurt, or one of the hundreds, if not thousands, of varieties of cheese that are made.

Think of the problems that an individual farmer would encounter in trying to convert a fat bullock, weighing up to 600 kg and producing 330 kg of bone-in beef, into something that could be cooked in an oven or frying pan. Even if it could be killed effectively and humanely, there would still be the problem of dealing with the less edible parts of the carcass, of preventing the meat from going rotten before it could be eaten, and of separating the various parts of the carcass into joints that respond best to different kinds of cooking. That is why butchery has been a skilled and separate trade from farming for hundreds of years, and why people have developed a range of food preservation methods, from the traditional drying, salting, smoking, and pickling to the more recent chilling, freezing, and even radiation treatment.

Farming therefore needs to be seen as part of the food chain, with many farmers producing a relatively small range of basic products that are then transformed into the enormous range of dishes that we put on the table either in the household kitchen or by the food processing and distribution industry. The crops accounting for the greatest sown area and output are listed in Table 4, and the main animal products in Table 5.

Note that the production and yield figures given in Table 4 are fresh weights. This does not make much difference in the case of cereals and pulses, but is significant for roots and tubers. Whereas rice and wheat, in store, contain about 12 per cent water and 88 per cent dry matter, potatoes have only about 20 per cent dry matter. Taking the world average yield figures, the right-hand column in Table 4 shows the effects of these differences for some important crops. Some farmers, of course, will be able to produce more than the world average: most European farmers would not be happy unless they were producing twice the world average wheat yield, and the world record wheat yield, set in New Zealand in 2010, is 15.7 tonnes per hectare.

Table 4 shows only some of the major crops. It does not, for example, show the area of grassland, which, according to one calculation, covers 3,500 million hectares across the world, a figure which includes 35 million hectares devoted to specialist fodder crops such as alfalfa. The precise figure is a matter of definition, because there are large areas of rough grazing, savannah, etc. that are not enclosed and which farmers' animals share with a wide variety of wildlife. Table 4 concentrates on the crops which are grown for human and animal food, but there are also about 35 million hectares of fibre crops, more than 90 per cent of which are devoted to cotton.

Table 4 The principal crops: world area, production, and average yield

	Area (million ha)	Production (million tonnes)	Yield (tonnes per ha)	Dry matter yield (tonnes per ha)
Cereals				
Wheat	215	670	3.1	2.7
Maize	177	872	4.9	4.3
Rice	163	720	4.4	3.9
Barley	50	133	2.7	
Sorghum	38	57	1.4	1.1
Millet	32	30	0.9	0.7
Oats	10	21	2.2	
Rye	6	15	2.6	
Sugar crops				
Sugar cane	26	1832	70.2	7.9
Sugar beet	5	270	55.0	
Roots and tubers				
Cassava	20	263	12.9	4.9
Potatoes	19	365	18.9	3.8
Sweet potatoes	8	103	12.8	3.8
Yams	5	59	11.7	3.5
Oil crops				
Soybeans	105	242	2.3	

(*continued*)

Table 4 Continued

	Area (million ha)	Production (million tonnes)	Yield (tonnes per ha)	Dry matter yield (tonnes per ha)
Rapeseed	34	65	1.9	
Sunflower	25	37	1.5	
Oil palm fruit	17	250	14.5	
Olives	10	16	1.6	
Pulses				
Beans, dried	29	24	0.8	
Chickpeas	12	11	0.9	
Cowpeas	11	6	0.5	
Lentils	4	5	1.1	
Fruit/vegetables				
Grapes/vines	7	67	9.6	
Bananas	5	102	20.6	6.2
Apples	5	76	15.8	
Tomatoes	5	162	33.7	
Onions	4	83	19.7	
Beverages				
Coffee, green	10	9	0.8	
Cocoa, beans	10	5	0.5	
Tea	3	5	1.5	

FAO.FAOSTAT. Production quantities (FAO2015) Accessed 26/7 April 2014 <http://faostat3.fao.org/browse/Q/QC/E>.

Note that area and production figures are rounded up, but yield figures are as original. For area comparisons, Great Britain is 22.7 million ha, France 55 million ha, Kenya 58 million ha, and Mexico 198 million ha.

It is important to remember, too, that some crops have multiple uses. Maize (known as corn in North America) is a good example. We are familiar with the corn cobs, sold fresh as sweetcorn, or canned as corn kernels. Steamed and crushed they make flaked maize, used as livestock feed, and with the addition of various flavourings they become cornflakes at breakfast time. They also produce starch, oil, and corn syrup. The whole crop, cobs, stalks, and all, can be harvested, chopped, and made into silage, now one of the main winter feeds for cattle in the US and Europe. And most recently, in response to higher oil prices and a legislative requirement to increase the use of renewable resources, an increasing proportion of the crop in the US has been converted into ethanol for incorporation into gasoline. By 2010 more corn was being used to make ethanol than was used to feed livestock.

There is a similar story for soya beans, which contain up to 20 per cent oil and 40 per cent protein. They are crushed to separate the oil from the remainder of the bean, which as soya bean meal is used as a livestock feed. It is the most widely used edible oil in the US (usually labelled as 'vegetable oil') and forms the basis of margarine, mayonnaise and other salad dressings, various baked products, and numerous other uses, but by 2011 nearly a quarter of all the soya bean oil produced in the US was converted into biodiesel. Similar stories could be told about other crops; farmers are not only part of the food chain, but of many other product chains too.

Livestock farmers also use a relatively narrow range of animals (see Table 2) to produce a much wider range of food and other products. Those accounting for the greatest volume of production are listed in Table 5. Comparing this with Table 2 we can see that the much greater number of chickens than any other form of livestock does not translate into greater meat production, simply because of their size.

Perhaps more surprisingly, although there are more cattle than pigs in the world at any one time, and the cattle produce two or

Table 5 Major livestock products, world total 2012 ('000 tonnes)

Meat		Milk	
Pigs	108,507	Cows	626,184
Chickens	92,731	Buffalo	97,942
Cattle	62,737	Goats	18,002
Sheep	8,481	Sheep	10,010
Turkeys	5,634	Camels	3,001
Goats	5,294		
Ducks	4,358	*Eggs*	
Buffalo	3,594	Hens	66,293
Geese	2,798		
Horses	766	*Wool*	2,067
Camels	511		

FAO.FAOSTAT. Livestock primary (FAO2015) Accessed 9 October 2014 <http://faostat3.fao.org/faostat-gateway/go/to/browse/Q/QL/E>.

three times as much meat per carcass, pigmeat (i.e. pork, bacon, ham, and all the various forms of preserved pigmeat) production is nearly twice as great as beef and veal output because pigs are ready for slaughter (at about six months) more quickly than cattle, which take at least twelve months for small cereal-fed animals, and more commonly eighteen to twenty-four months for grass-fed cattle. And of course quite a lot of the cattle population are dairy cows which would be unlikely to be slaughtered before their fifth year, and could last much longer.

China is the biggest producer of pig, chicken, sheep, and goat meat, the US of beef and turkey meat, and India and Pakistan dominate buffalo production. Camel meat is the only product in which African countries dominate the market, with Sudan, Somalia, Kenya, and Egypt among the larger producers.

Over the twenty years between 1992 and 2012 the total world output of many of these animal products increased significantly: chicken doubled, eggs nearly doubled, goat meat production increased by about 85 per cent, and pig production by about 50 per cent. Despite these increases, roughly four billion people rely on rice, maize, or wheat as their staple food. Another 500 million rely on cassava (manioc), a root crop that originated in South America but is now widely grown and eaten in sub-Saharan Africa. It is drought-resistant, so it is often grown as a 'famine crop', to be left in the ground and harvested when other staples are scarce. In some African and Asian countries such as Mali, Niger, Bangladesh, and Cambodia, cereals supply over 70 per cent of dietary energy, whereas in North America and Western Europe the comparable figure is less than 30 per cent.

Most people in these richer countries have a diet that supplements the carbohydrates supplied by cereals and potatoes with proteins and fats from meat and dairy products. But of course the animals that produce these products have to be fed, and while some of their feed comes from grass and other high-fibre plants that people cannot digest, some comes from cereals and vegetable protein-producing crops such as soya beans. Over 70 per cent of the cereals grown in Canada in 2005, and 40 per cent of the wheat grown in Britain, went to feed animals. On a world scale, it is estimated that about a third of all cereals are fed to animals. Soya bean production in Argentina and Brazil, taken together, nearly quadrupled between 1993 and 2013, as their cultivation expanded into areas previously used for grazing or covered in forest, and about 80 per cent of the world's soya bean harvest is used for animal feed.

Does this mean that Argentina and Brazil have dramatically increased the number of animals that they produce, and are consequently consuming much more meat and more dairy products? Not necessarily, because, as we saw earlier, it is important to see agriculture as a producer of raw materials for the food industry, or as the first link in the food chain. Thus any farm product might be consumed on the farm, within a local market area, perhaps in the country in which it was produced, or in some other part of the world. In order to understand why this happens we need to know more about how agricultural markets work, and that is what we consider in the next section of this chapter.

The market for farm products

The county of Kent in south-eastern England is traditionally noted for the extent and quality of its fruit farms. In the autumn of 2014 apple growers there were expecting a bumper harvest. The previous winter had been cold, keeping pests at bay, there was enough rain in August to help the fruit to grow, and a beautiful sunny September enabled picking to begin early. It was much the same in the rest of Europe. According to newspaper reports, one farmer couldn't remember a better growing year, and another spoke of getting 'quite a buzz, seeing a fantastic crop'. But they both felt that their apples would be worth little, and one was preparing to make a loss while the other felt that he might just cover his costs. This seems crazy, but it's not a new problem. The Porter in Shakespeare's *Macbeth* imagines himself opening the gates of hell to 'a farmer that hang'd himself on th' expectation of plenty'. So how can a wonderful crop lead to a financial loss? If we look at the way in which agricultural markets work we should be able to explain it.

As with any other market, agricultural markets are affected by changes in demand and supply. If demand increases but the supply remains unchanged, prices are likely to rise; if supplies increase but demand remains constant they are likely to fall. We therefore need to know what affects the demand for and supply of farm products.

The demand for a product is influenced by its price, the incomes and tastes of the potential consumers, and the number of those consumers. When prices are high consumers think hard before buying a product, and when they are low they consider buying more of it. That's the basic idea, but of course it also depends upon what the product is.

For staple foods, such as rice, maize meal, pasta, bread, or potatoes, the demand will not change much. When prices are high people still need the basics of their diet, so they will pay what they have to. And when prices are low they are not likely to consume vastly increased quantities of their staple foods, but instead use the money that they have saved on some other product. Conversely, the market for luxury foods will be much more affected by the price level, because they are not a vital part of the diet, so a small price increase might cause a big decrease in the quantity consumed, and vice versa. The possibility of substitution has an impact here too. We can obtain protein from the more expensive cuts of meat, which are tender and taste good, but we can also get it from cheaper cuts. Similarly, high beef prices might increase the demand for pork, as long as consumers see roast pork as an acceptable substitute for roast beef. However, some consumers may have cultural or religious objections to eating pigmeat in any form, so their preferences would not affect the pork market.

This is just one of many examples of the way in which consumer tastes affect demand. Governments in many industrialized countries have attempted to influence consumer food preferences for health reasons, encouraging, for example, the consumption of less red meat and more fruit and vegetables. Whether these campaigns have been as successful as commercial advertising, which is the other main influence on consumer tastes, is a matter of controversy (which is further discussed in the Very Short Introductions on *Food* and *Nutrition*).

Income levels affect the demand for food in all sorts of ways. A study carried out for the United States Department of Agriculture in the early 21st century compared the behaviour of consumers in low-income countries, with incomes less than 15 per cent of those in the US, and high-income countries, with incomes of 50 per cent or more of the US level. Some of the results of the study are given in Table 6. They show that people in richer countries spend a smaller proportion of their incomes on food than those in poorer countries, and that their food consumption behaviour is less responsive to income changes. As people get richer they spend

Table 6 Income effects on the demand for food

	Low-income countries	High-income countries
Percentage of budget spent on food	47	13
Increase in demand for food when income increases by 1%	0.73%	0.29%
Percentage of food budget spent on cereals	28	16
Increase in cereal consumption when income increases by 1%	0.56%	0.19%
Percentage of food budget spent on meat	18	25
Increase in meat consumption when income increases by 1%	0.82%	0.33%
Percentage of food budget spent on dairy products	9	14
Increase in consumption of dairy products when income increases by 1%	0.93%	0.35%

A. Regmi, M. S. Deepak, J. L. Seale Jr, and J. Bernstein, *Cross Country Analysis of Food Consumption Patterns*, Economic Research Service, US Department of Agriculture., tables B-1 and B-2, available at <http://www.ers.usda.gov/media/293593/wrs011d_1_.pdf> (accessed 20 October 2014).

more on foods containing proteins and fats, like meat and dairy products, but consumption of staple foods such as cereals increases less rapidly. The study also found that, perhaps not surprisingly, the poor are more responsive to food price changes than the rich.

These data are bad news for farmers in rich countries, because they imply that the demand for food will not increase much unless the population increases, and population increases in rich countries tend to be lower than in poor countries. Conversely, they are bad news for consumers in poor countries, because they imply that small increases in food prices or small decreases in incomes will have big effects on the amount of food that they can buy.

Does a slow demand increase matter to rich-country farmers? It does if they increase supplies more than demand, because the prices they receive will then fall. In theory, this should be a self-regulating system, because lower prices should result in less being supplied to the market, as all farmers reduce their output, or some go out of business altogether, and supplies thus fall to the level at which they equate to the quantity demanded.

In practice it's more complicated. Product prices are not the only influence on farm output. Production costs can change: pig and poultry profitability, for example, is very sensitive to changes in the price of feedingstuffs. Farmers can adopt new varieties of crops, or breeds of livestock, which produce more without needing more inputs, or any of the other myriad technical changes that increase output or reduce costs. Government policies can change, so that farmers are encouraged to produce more, as they were after World War II and for many years after in Europe, or less, as they were when this policy succeeded to the point of producing surpluses.

It is also important to remember that agricultural production cannot be turned on and off at the flick of a switch. If farmers decide that they want to produce more milk it could take them more than two years to raise heifer calves to the point at which they are ready to be mated, become pregnant, deliver a calf, and take their place in the dairy herd. Tree crops may also take years to come into production. Overall, therefore, with some exceptions, output of many farm products tends to respond only slowly to changes in market conditions.

The reason why apple growers in Kent might have mixed feelings about a bumper harvest should now be clear. The demand for apples in the UK is falling as the range of fruit sold in supermarkets increases, and falling prices will not have much impact on the quantity of apples that consumers buy. In addition, UK farmers have to compete with suppliers in other countries who also sell their apples into the UK market, and that takes us on to another important issue for farmers: the question of international trade.

International trade in farm products

As we saw earlier, some farmers simply produce what they and their immediate families want to eat, and sell none of their output. They are subsistence farmers, and there are not many of them. Most farmers probably sell some of their produce, simply because they need things that they cannot produce for themselves, from salt and iron to fertilizers and pesticides. Farmers have sold food and fibres to non-farmers almost since the beginning of farming. Eventually this led to regional specialization, as farmers on good arable soils realized that they could make more money by specializing in crop production, and those on poor upland farms realized that they might do better by exchanging their livestock for the crops they needed. The logical end result of this process is that countries specializing in agriculture should export farm products in exchange for the manufactured goods produced in industrialized countries.

This was the pattern that had emerged by the end of the 19th century. The bulk of international trade then was in temperate products, sent from countries that had been settled by European emigrants, such as the US, Canada, Argentina, Australia, and New Zealand, back to Europe, or from Eastern Europe and Russia to Western Europe. Tropical products, such as tea, coffee, cocoa, and oilseeds, accounted for the minor part of the total.

Table 7 The top ten importers and exporters of agricultural products in 2012

Exporters	% of world exports	Importers	% of world imports
EU 27	37.0	EU 27	35.7
(EU exports to non-EU countries)	(9.8)	(EU imports from non-EU countries)	(9.9)
US	10.4	China	9.0
Brazil	5.2	US	8.1
China	4.0	Japan	5.4
Canada	3.8	Russia	2.4
Indonesia	2.7	Canada	2.2
Argentina	2.6	Republic of Korea	1.9
India	2.6	Saudi Arabia	1.7
Thailand	2.5	Mexico	1.6
Australia	2.3	India	1.5
% of world total (by value) exported by the top 10	73.1	% of world total (by value) imported by the top 10	69.5

World Trade Organization, *International Trade Statistics 2013, part II, Merchandise Trade*, © World Trade Organization, 2013, <http://www.wto.org/english/res_e/statis_e/its2013_e/its13_merch_trade_product_e.pdf, table II.15> (accessed October 2014).

Perhaps surprisingly, there are still echoes of this in the present pattern of world agricultural trade. It continues to be dominated by industrialized countries, as Table 7 shows, although some newly industrializing countries have now joined the list of major importers. Europe is still the major importer, and the US, Canada, Argentina, and Australia remain in the top ten exporting countries. The EU, the US, and China appear as both importers and exporters simply because they are major importers of some commodities and major exporters of others. The US, for example, is one of the top five exporters of wheat, rice, maize, and beef, and among the top five importers of sugar, coffee, cocoa, and tea. There is considerable trade among the EU countries, which accounts for the EU's position at the top of both lists, and traders in some EU countries act as both importers and exporters: Germany, for example, is a major importer and exporter of butter and cheese.

Table 8 World exports as a percentage of world production in 2011–12

Wheat	22	Pigmeat	13
Maize	13	Chicken	13
Rice	5	Beef	16
Sugar	22	Sheepmeat	10
Potatoes	3		
Bananas	18	Milk (equivalent)	14
Soyabeans	38		
Coffee	84	Eggs	3
Cocoa	64		
Tea	40		

FAO.FAOSTAT. Crop and livestock products. Exports of selected countries (FAO 2015) Accessed 10 October 2014 <http://faostat3.fao.org/browse/T/TP/E>.

Although the bulk of the tropical beverages—tea, coffee, and cocoa—are produced for export, and a significant proportion of the soya beans are exported, trade in foods generally represents only a small proportion, often less than 20 per cent, of world production, as Table 8 demonstrates. When harvests are poor, domestic requirements tend to be satisfied first, so less is available for export, whereas much of the surplus from good harvests will be sent to export markets. Thus the supplies available to the world market tend to fluctuate more than world output, and world market prices are correspondingly variable.

For example, consider the case of a hypothetical country exporting the world average percentage of rice from a harvest of 100 million tonnes. In a normal year it will supply 5 million tonnes to the world market. In a good year when production increases by, say, 5 per cent, it will have 10 million tonnes available for sale, and in a poor year, with output 5 per cent down, none at all for export if its domestic requirements are to be satisfied.

Governments are understandably concerned about these questions of international trade and domestic demand and supply. Even in industrialized countries with varied and plentiful supplies, food prices can be a sensitive political issue; in poor countries the supply/demand balance can be a matter of life and death. Thus most countries have some kind of government involvement in agriculture, which may range from statistical monitoring of inputs and outputs, or provision of advice to farmers, to comprehensive control of production, prices, and farm incomes. Even governments in those industrialized countries that favour the operation of market forces, such as those of the EU or the US, take measures to support farm incomes and affect the trade in farm products.

At an international level, the World Trade Organization monitors agricultural trade and through periodic intergovernmental meetings attempts to reduce barriers to the free flow of trade. The

Food and Agriculture Organization of the United Nations collects a comprehensive range of agricultural statistics and produces numerous reports on the state of world agriculture and what can be done to address its problems, especially in developing countries.

The purpose of this chapter has been to show that the output of the agricultural industry, with all its variety, is simply a part of the complex of local, regional, national, and international links that have emerged to try to ensure that people have something to eat every day. These linkages also extend to the other side of the agricultural industry, to provide the inputs that farmers need, and they form the subject of Chapter 4.

Chapter 4
Inputs into agriculture

The story is told of an old farmer who took over some land that had not been farmed for years. It was overgrown with weeds, the gates were hanging off their hinges, the hedges were either overgrown or full of gaps, the drains and ditches were blocked so that the lower fields were boggy. After several years of hard work and considerable investment the land was transformed. There were flourishing crops growing in well-drained fields with not a weed to be seen. The pastures, now surrounded by trimmed stockproof hedges and new gates, fed handsome cattle and sheep in great quantities. One day, as the farmer was leaning on the gate admiring it all, the parson came by. 'You and the Good Lord have made a wonderful farm here, John', said the parson. 'Perhaps so', John replied, 'but you should have seen it when he had it to himself'.

Left to itself, land will produce food for people, as Palaeolithic hunter-gatherers knew, but their population densities remained low. To feed the present human population, farmers use the land and add to it their labour, their animals, buildings, tools, and machinery, a variety of products bought in from other industries, from fertilizers and pesticides to computers and mobile phones, and professional services from vets, advisers, and scientists. This chapter is about those inputs. But first, the land.

Land

It is possible to produce food without land. Crops can grow in a nutrient solution, a technique called hydroponics. Increasingly salad crops are grown in greenhouses, in which large quantities of produce come from relatively small areas of land. Intensive livestock, such as pigs and poultry, are also housed on relatively small areas of land, although of course much greater areas of land are needed to produce their feed. But for most farmers, land, 'the original and indestructible powers of the soil' in the British economist David Ricardo's words, is a basic necessity.

The world's land area is just over 13 billion hectares, of which 37.6 per cent, or nearly 5 billion hectares, is agricultural land. The rest is approximately equally split between forest land and other uses, which range from urban land to steep uncultivable mountains and land covered in ice. Of the agricultural land, 28.3 per cent is arable and 3.1 per cent in permanent crops, such as coffee or apple trees. The rest is meadows and pastures, of varying kinds and qualities.

The great bulk of the world's food comes from its agricultural land, which is far from evenly distributed. Only 44 per cent of the land in the US, one of the world's great agricultural exporters, is in agriculture, compared with over 70 per cent in Bangladesh but only 12.6 per cent in Japan. There are similar variations in the amount of cropland per head of population, from 2 hectares in Australia to 0.03 hectares in Japan. The world average in 2010 was 0.22 hectares per person, which is half of what it was in 1960. This means that food output per hectare must have grown over the last fifty years, which indeed it has, by between 2 and 4 per cent per year. At the same time the amount of cultivated land has also increased, by about 1 per cent per year, but the potentially cultivated land which is not yet farmed is very unevenly distributed. According to the Food and Agriculture Organization

(FAO) of the United Nations, there is no 'spare' land in southern or western Asia or North Africa, and 90 per cent of the accessible land is in seven countries: Congo, Angola, Sudan, Brazil, Argentina, Colombia, and Bolivia. Since its further accessibility would have environmental and social implications, we might question whether there is any spare land at all.

It should be clear from Chapter 1 that not all land is equally productive. Variations in soil, altitude, aspect (whether a slope is north- or south-facing—south-facing slopes get more solar radiation in the northern hemisphere and vice versa), and climate, especially as regards temperature and rainfall, will all affect productivity.

For example, in global terms, Tanzania is well-provided with agricultural land, with 0.77 hectare per person, but much of this land is dry savannah, suitable only for extensive grazing. Japan, in contrast, as we have seen, has very little agricultural land, but much of it is intensively cultivated to produce rice and vegetables. While Tanzania has more than twenty times as much land per person as Japan, the value added per Japanese hectare is sixty-seven times the value added on a Tanzanian hectare. Hence the idea of grading land according to its flexibility: the best land can produce more or less anything, and as land quality decreases there are increasing limits on what it can produce.

Water is often the limiting factor. In some cases the land is too wet, or wet at the wrong time of year, and needs to be drained. More commonly it is too dry, and requires irrigation. Some crops, such as certain types of rice, are entirely dependent upon irrigation, and crop production in drier parts of the world can also be reliant on irrigation.

There are well-known examples in the Central Valley of California and in the High Plains, that drought-prone area to the east of the

Rocky Mountains, part of which is underlain by the Ogallala aquifer, which extends from South Dakota to north-west Texas. Its use illustrates both the potential and pitfalls of irrigation. The development of high-volume pumps and centre pivot irrigation systems enabled the area of land irrigated from the aquifer to increase from about 800,000 hectares to over five million hectares between 1950 and 1980, and the resulting green crop circles are still clearly visible on satellite photographs of the region (see also Figure 4). By the late 1970s, what had been a drought-stricken dustbowl was producing corn, cotton, wheat, and sorghum, but by then it was clear that the Ogallala aquifer was being emptied far faster than it was being filled, and, as also happened in parts of California, the water table was dropping.

There are other examples of unsustainable extraction on the plains of northern China and the Indo-Gangetic plain in India. This has led to the concept of the 'water footprint', analogous to the carbon footprint: the idea that different crop and animal products require very different quantities of water in order to provide the nutrients that we derive from them. Most (78 per cent according to one calculation) of the water used in global crop production is rainwater, which is called 'green water' in these calculations. 'Blue water' is the surface and groundwater, which either evaporates, flows to the sea, or is used in agriculture. It provides about 12 per cent of the water used in crop production. The third category is 'grey water', which is the volume of water needed to assimilate any pollutants, and it accounts for the remaining 10 per cent of the water use.

Farm animals too use water, for drinking, obviously, but a lot is also used in cleaning their housing and providing the food they eat. Overall, the agricultural sector accounts for about 85 per cent of global blue water consumption. Table 9 shows the considerable variation in water footprint of various sources of energy, proteins, and fats.

Table 9 Water footprint per unit of nutrition

Energy (litres per kcal)	Protein (litres per gramme)	Fat (litres per gramme)
Beef 10.19	Nuts 139	Starchy roots 226
Sheep/goat meat 4.25	Beef 112	Pulses 180
Chicken meat 3.00	Pigmeat 57	Beef 153
Sugar crops 0.69	Eggs 29	Milk 33
Cereals 0.51	Pulses 19	Pigmeat 23
Starchy roots 0.47	Oilcrops 16	Oilcrops 11

Data from M. M. Mekonnen and A. Y. Hoekstra, 'A Global Assessment of the Water Footprint of Farm Animal Products', *Ecosystems*, 15, 2012, pp. 401–15.

Labour

For most of human history, the majority of people lived outside towns and cities. But since 2008 the world's urban population has been bigger than the rural population, and it is predicted that urban families will account for the majority of future population growth. This obviously has implications for agriculture, because the majority of agricultural workers live in the countryside, although it must be remembered that a strict distinction between agricultural and non-agricultural workers is more of a first-world idea than something that would be recognized in developing countries, where many city dwellers—perhaps over 40 per cent of the population of Ouagadougou, the capital of Burkina Faso, for example—are still involved in the production of crops and livestock.

Taking the world as a whole, over a quarter of the population (i.e. the economically active population and their non-working dependants) is employed in agriculture, but within this total there are great variations from one country to another. Whereas the figure for Africa as a whole was 49.1 per cent in 2010, and for India it was 48.4 per cent, in Latin America it was 15.8 per cent,

6. Farmers in Somoa village, upper west Ghana, sell their farm products, including millet (on the right of the picture), at market.

and in the US, Canada, and the UK it was only about 1.8 per cent of the population. However, quoting these international statistics raises a question that is not as easy to answer as it might seem at first sight: who should be included in the category of agricultural worker, and what counts as a farm?

Throughout this book we have referred to the people carrying out agricultural operations as 'farmers', but this is the point at which we should recognize that we have used the term to cover the activities of a wide range of people. It is certainly true that most farms in most countries are owned by an individual or group of individuals and use mostly household labour (see Figure 6). In other words, they are family farms.

Having said that, there are other people who are involved in agriculture without being members of family farming households. Most obviously, there are those who work on a farm without having a share in its ownership. They are variously called farm workers or agricultural labourers and they work for wages. Conversely, there are those who own farm land but do not work it themselves. They are landowners, and they lease their land to tenant farmers in return for a rent. Sometimes governments, or religious institutions, or corporations act as landlords. Alternatively, corporations (sometimes referred to as 'agribusinesses') or co-operatives may employ managers and workers to farm the land rather than leasing it to family farmers. It would be quite possible to find farms on which a family owned some of the land, rented some more, and also had rights to use other land in common with other farmers, for grazing, for example.

Over time, these different roles have often acquired labels with implied status. Thus small family farmers are often called peasants. In some countries the term has connotations of taking pride in independence and self-sufficiency; in others it may imply a low status and lack of sophistication. Landowners, who do not work the land, are sometimes called landlord, a word which by itself seems to imply high status. Conversely, the implicit low status ascribed to landless agricultural labourers covers people who may be poorly educated, low-paid migrant workers, or, in some developed countries, highly trained and well-paid agricultural graduates.

As we might expect, if there are many kinds of farmers, there are also many definitions of what constitutes a farm. At first sight it seems pretty obvious that a farm is an area of land, with its associated buildings and machinery, on which people grow crops and raise livestock. But which people, how much land, and what sort of crops or livestock?

Countries vary in what they consider as the minimum size for a farm when responding to the World Census of Agriculture, which is carried out by every decade by the UN's Food and Agriculture

Organization. India has no minimum, whereas Bangladesh only included holdings of greater than 0.2 hectares, and China included farms of only 0.07 hectares. Germany has a minimum farm size of 2 hectares. In Russia, 98 per cent of the holdings are owned by private individuals, but they only cover 2 per cent of the farmland area; the other 98 per cent of the farmland is owned by institutions, entrepreneurs, and non-profit citizen associations.

Thus any figure for the number of farms in the world cannot be exact, because it depends upon definition, but the latest attempt to estimate it concluded that there are now more than 570 million farms in the world, and that the number is increasing as a result of growing numbers in low- and middle-income countries. More than half of all farms are in China and India; only 4 per cent are in high-income countries. The bulk of them—more than 500 million—are family farms, in the sense that the family owns them and does most of the work on them, and most of them are small: 94 per cent of all the farms in the world have less than 5 hectares of land.

However, looking at farm sizes through world averages hides a lot of geographical variation. In East and South Asia, and in sub-Saharan Africa, most of the farms are of less than 5 hectares. They account for at least half of the agricultural area, whereas in Latin America there are fewer very small farms and most of the land is in a few of the larger farms. In high-income countries, although many small farms remain, the bulk of the land is worked by the big farms. In the US, for example, 80 per cent of the output, in value terms, is produced by the biggest 10 per cent of farms. Table 10 gives some examples from individual countries.

Table 10 illustrates one of the difficulties in discussing farm size: if you ask farmers about the size of their farms they will almost always answer in terms of hectares or acres or whatever the local measure of land area is, or perhaps in terms of the number of livestock they have, but the question is often really about the size

Table 10 The distribution of farms and land in the late 20th century

	Ethiopia	Ecuador	UK	US
Number of holdings ('000)	10,758	843	233	2,129
Total agricultural area ('000 ha)	11,047	12,356	16,528	379,712
Holding size (ha) trend	Falling 1.4 → 1.0	Falling 15.3 → 14.7	Rising 55 → 71	Rising 158 → 178
% of holdings of less than 5 ha	99	64	23	11
% of land in holdings of less than 5 ha	93	6	1	0.14
% of holdings of 5–100 ha	1	35	60	38
% of land in 5–100 ha holdings	7	65	30	12
% of holdings with more than 100 ha	0	1	17	27
% of land in holdings with more than 100 ha	0	29	69	88
Average number of hired permanent workers per farm	0	0.3	0.6	1.4

Data from *Food and Agriculture Organization of the United Nations*, 2014, S. K. Lowder, J. Skoet, and S. Singh, 'What Do We Really Know about the Number and Distribution of Farms and Family Farms Worldwide?', background paper for *The State of Food and Agriculture 2014*, ESA Working Paper No. 14–02, Rome: FAO <http://www.fao.org/docrep/019/i3729e/i3729e.pdf>. Reproduced with permission.

of the farm business, and whether or not it will support the farm family. A small farm on very fertile soil with intensive cropping may produce far more than a much bigger farm, in terms of land area, on poor land suitable only for grazing a few cattle.

However, international statistical organizations collect farm size data in area terms only, so it is those that we have to use. And clearly the fact that 29 per cent of the agricultural land in Ecuador is in the hands of the 1 per cent of holdings with more than 100 hectares tells us that a significant proportion of the country's agricultural output is likely to come from those farms. Similarly, in the UK, the vast majority of the farms have less than 100 hectares of land, but, as in the US, the bulk of the output comes from those with more than 100 hectares.

This often has important political implications. An agricultural policy designed to help big farms may make life difficult for small farmers. However, each of the smaller farmers has a vote, both in government elections and in elections for office-holders in farmers' pressure groups, so their concerns may achieve more political prominence than those of the bigger farmers, despite the fact that the latter may have more spare time and money to devote to political activities. On the other hand, even the biggest farms have less market power than food processors and retailers in developed countries, so they have to accept the prices that they are offered just as the smaller farmers do.

The other obvious point illustrated by Table 10 is that small farms are more likely to rely on family labour and less likely to hire labour than big farms. As the table shows, while farm sizes are increasing in the UK and the US, and in many other developed countries, in developing countries they are often decreasing as populations increase. This means that families may have too little land to use their available labour efficiently. If there are local employment opportunities in the non-farm rural economy that may be a positive feature, but otherwise it may result in permanent or temporary migration to urban areas in search of work.

Examined from the viewpoint of international statistics, and perhaps even from the perspective of national agricultural policymakers, the relationship between land and the labour that works it is complex enough, but it is worth remembering that we are talking about families that don't just work the land—they usually live and work on *their* land, a particular bit of land, which they may have worked for a long time, and which their parents and grandparents may have worked before them. Thus they have a connection with it that may go beyond the purely economic.

European farmers, asked about their business objectives, may talk about profit maximization or long-term growth, but many of them will also say that their principal purpose is to hand on a viable farming business to the next generation, just as one was given to them. Exactly the same view was articulated by a representative of Maasai pastoralists in Tanzania when the government reportedly wished to turn their land into a game-hunting reserve. They were offered financial compensation, but 'you cannot compare that with land. It's inherited. Their mothers and grandmothers are buried in that land. There's nothing you can compare with it.' Whether this will continue as family farms grow bigger, and more land in high-income countries is farmed by farming corporations, remains to be seen.

Increasingly, farmers and farm workers are assisted by numerous other specialists. Originally, perhaps, farmers made all they needed from the resources of the farm, with assistance from the blacksmith in making ploughshares, and from the miller in processing their grain. Modern farmers may have their fields ploughed and sown and reaped and mown by specialist contractors operating big machines. When their own machinery breaks down they can call for a mechanic from the local machinery dealer. Their cattle may be artificially inseminated by a travelling specialist who visits the farm, and their fat cattle, sheep, and pigs will be driven to the slaughterhouse by a lorry driver from a livestock haulage firm. An agronomist may tell them which pesticide to use on their crops, and consultants, advisers, and bank managers may discuss their business plans. New technology may be developed for them by agricultural scientists working in universities and research institutes or for commercial firms. They may be made aware of it when they are trained in agricultural colleges and universities, and subsequently by agricultural journalists and broadcasters, or by the sales staff of firms in the agrochemical and farm machinery industries.

After leaving the farm, their crops and livestock will be passed on to the food processing and retailing industries, so although the numbers directly involved in agriculture in developed countries may be a tiny proportion of the labour force, many more people are involved in the food chain as a whole.

Other inputs

When people first began to cultivate plants to eat, as opposed to simply collecting what they found growing wild, they would have needed some kind of digging implement to make a seedbed, sickles to cut their cereals, and baskets in which to collect them. But much of the work, and the power behind it, would have come from human hands and muscles. Over time, the proportion of the work done directly by the human hand has diminished, and the importance of the other things that farmers use has increased.

The list is almost endless, from draught animals and breeding flocks and herds, to fruit trees, buildings, fences, hedges, ditches, drains, tools, machinery, fertilizers, and pesticides. They are the things that farmers use in the course of producing crops, animals, and animal products. Most of these were, at one time, produced on the farm. A harrow to rake over a seedbed could be as simple as a bush cut from a hedge, pulled by an ox or a cow. The cows in the dairy herd and the sows in the pig herd were bred on the farm. The wood to make fences or the framework of buildings could be cut from nearby woodland or the farm's hedges. The hedges themselves started as young cuttings planted by the farm staff, who also dug the ditches beside them. The only major fertilizer was the manure produced by the farm's animals, although in some places farmers might be fortunate enough to live within reach of some other nutrient source, such as seaweed. Weeds were controlled by hoeing. And in some developing countries this continues. But in developed countries, modern equipment has largely taken over. Perhaps the most iconic item is the tractor, which has almost completely replaced draught cattle and horses in developed countries.

Different kinds of farm demand different kinds of input. Land and labour will account for a relatively greater proportion of the total costs on an extensive grazing farm on which the farm workers spend most of their time looking after the animals and the land receives little fertilizer, whereas on an intensive highly mechanized arable farm the machinery, pesticides, and fertilizers will account for more of the expenditure. Table 11 gives some idea of the enormous variability between different countries in the use of some of those inputs that are more easily quantified. It shows that richer countries tend to use more purchased inputs than poorer countries, which is what we would expect, but also that input use also varies with the intensity with which the land is farmed.

These huge variations in machinery, pesticide, and fertilizer use (and in other inputs, less easily quantified, such as buildings) mean that the carbon footprint of agriculture varies from one country to another, just as it varies from one crop or livestock product to another, or one food consumer to another.

There are three principal greenhouse gases: carbon dioxide (CO_2), methane (CH_4), which has twenty-three times the global warming potential of CO_2, and nitrous oxide (N_2O), with 296 times the global warming potential of CO_2. Carbon dioxide is produced by the internal combustion engines powering farm machinery, in the manufacture of fertilizers, especially nitrogen fertilizers, and when land is cleared from forest or ploughed. Ruminant animals emit methane as one of the by-products of their digestive processes,

Table 11 Purchased inputs in agriculture in various countries, 2000–10

	Tractors per thousand ha of arable land	Tractors per thousand workers in agriculture	Pesticide use (kg per ha)	Fertilizer use (total of N, P, and K, kg per ha)
Mali	0.2	0.1	0	7.4
Tanzania	2.1	0.8	No data	7.5
Ecuador	4.0	4.6	3.7	88.2
Brazil	12.9	28.4	No data	111.7
India	13.2	3.7	0.2	156.6
US	27.0	707.1	2.2	107.7
France	64.1	603.6	2.9	140.2
UK	75.6	976.4	3.0	238.2
Japan	472.0	420.6	13.1	271.1

FAO, *FAO Statistical Yearbook 2013: World Food and Agriculture*, Rome: FAO, 2013. Reproduced with permission.

and nitrous oxide is produced from the breakdown of fertilizers and animal manure. Thus highly mechanized agricultural industries using lots of fertilizer produce more greenhouse gases than less developed agricultures, but even low-intensity cattle grazing will have some impact on global warming through the production of methane.

There are widely varying estimates of agriculture's contribution to greenhouse gases, ranging from 17 to 32 per cent of the total. A Food and Agriculture Organization study found that the world's livestock alone accounted for 9 per cent of carbon dioxide, between 35 and 40 per cent of methane, and 65 per cent of the nitrous oxide in global anthropogenic emissions. Different foods similarly have widely varying carbon footprints, as Table 12 shows,

Table 12 UK greenhouse gas emissions (kg of CO_2 equivalent per kg of product)

Beef	68.8	Coffee	10.1
Mutton	64.2	Rice	3.9
Animal fats	40.1	Sunflower oil	3.3
Pigmeat	7.9	Wheat	1.0
Poultrymeat	5.4	Beans	0.8
Eggs	4.9	Potatoes	0.4
Milk	1.8	Sugar	0.1

Source: Data from P. Scarborough, P. N. Appleby, A. Mizdrak, A. D. M. Briggs, R. C. Travis, K. E. Bradbury, and T. J. Key, 'Dietary Greenhouse Gas Emissions of Meat-Eaters, Fish-Eaters, Vegetarians and Vegans in the UK', *Climate Change* 125, 2014, pp. 179–92.

Note that much lower figures for Norway are shown in D. Blandford, I. Gaasland, and E. Vårdal, 'Extensification versus Intensification in Reducing Greenhouse Gas Emissions in Agriculture: Insights from Norway', *Eurochoices* 12(3), 2013, pp. 4–8.

with animal products and imported commodities having higher figures than plant products.

Thus the carbon footprints of individual consumers vary with their diets. A UK study found that those eating more than 100 grammes of meat per day (about two chipolata sausages plus two slices of roast beef) would be associated with emissions of 2,624 kg of CO_2 equivalent per year, whereas the figure for vegetarians was only 1,391 kg. (For comparison, the carbon footprint of a ten-year-old small family car covering 6,000 miles per year was 2,440 kg.) Will both farmers and food consumers come under pressure to reduce these carbon emissions in the future?

It is important to remember that farmers do not just use *things* to help them to cultivate the land and care for their animals. Tradition, experience, and knowledge are important too. Knowing when a crop or an animal is doing well, or spotting problems in time to do something about them, are an important part of a farmer's everyday job. Many farmers acquire ideas about the importance of good livestock care or 'doing right by the land' as part of their upbringing or training in farming. An economist would say that it was part of the social and intellectual capital of the industry. As with the other kinds of capital, such as machinery and buildings, it can be increased by investment, in this case in education and training or research and development. Many countries now have at least one, and in some cases a network, of agricultural colleges and university departments of agriculture, offering both full-time courses and other training opportunities. Many of them also carry out research, or are linked to other research institutions.

Whereas roughly fifty years ago most of the world's agricultural research was carried out in developed countries, the present pattern is that an increasing proportion of research spending is found in the newly industrializing countries. The combined agricultural R&D spending of China, India, and Brazil in 2008 was $US7.6 billion, compared with a total of $US9.2 billion in the US, Japan, and France. It is probably too soon to say how this will affect farming in those countries, and in developing countries. The experience of the agricultural industries in North America and Western Europe suggests that it takes time for effective knowledge networks to develop and transmit new technologies.

Farming systems

If we now bring together the information about the outputs of farming from Chapter 3 with the data on inputs that we have compiled in this chapter, we should be able to see how differences can combine to produce different farming systems.

This is something that has fascinated some academic experts on agriculture for many years, and they have produced a variety of more or less complex analyses, coupled with the lessons that might be learned from them for purposes of farm management and agricultural policy making. The basic ideas, however, are fairly simple, and we have implicitly used them already. Each farm differs according to where it can be placed on several continuously varying scales: intensive/extensive, simple/advanced technology, large/small size, mixed/specialized, tropical/temperate, close to/far from markets, and so on. Chapters 5 and 6 will look at the ways in which these differences affect farming as it is presently carried out, and how farmers might deal with the problems they have to face in the future.

Chapter 5
Modern and traditional farming

Your ideas about the difference between modern and traditional farming probably depend on where you live, and when. People old enough to remember back to the 1950s who live in paddy rice-producing areas might think of traditional farming as involving tall crops cultivated with the help of animals using only organic manures. Where yams are a staple part of the diet they might think of crops grown on small hills or ridges produced by hand labour, with long fallow periods between crops.

Farming in McLean County, Illinois, about a hundred miles south-west of Chicago in the heart of the US's Corn Belt, was changing from one tradition to another in the 1950s. It was a time when tractors were replacing horses, which meant that the oats and hay grown for the horses could be replaced by more corn and soya beans, and about half of the farms had mechanical corn pickers. But mechanical cultivators and hoes were still the only way of controlling weeds, and farms were family businesses. Several hundred miles to the west, however, traditional farming meant cattle ranching, with much bigger landholdings, whereas a few hundred miles south it meant small sharecroppers growing cotton.

In Europe, in the early 1950s, traditional farming meant mixed farming on family farms. On a typical farm in lowland England, for example, there might be a small dairy herd with Shorthorn, Ayrshire, or Jersey cows, with the calves that were not needed as herd replacements reared for beef, and a sheep flock. A few pigs would be found somewhere in the farm buildings, and chickens and geese would be wandering round the farmyard. Out in the fields there would be cereal crops, often a few acres of potatoes, and sometimes sugar beet. A good proportion of the land, perhaps a quarter to a half, would be in grass, producing summer grazing for the cattle and sheep and a crop of hay for winter fodder. There would be a few acres of root crops such as swedes, turnips, or mangel wurzels, also for winter fodder.

Mechanization would not have gone as far as it might have done in McLean County, but the cows would usually be machine milked into a bucket that was carried around the cowshed. Small tractors such as the little grey Ferguson were becoming more and more popular, and combine harvesters were beginning to replace the reaper-binder. But it is important to remember that this is the image of traditional farming in the mind of somebody who was young in the 1950s. Middle-aged farmers at the time would have thought of horses and hand milking as traditional, and tractors and milking machines as modernity. And if we go back a hundred years before they were born, to the early 19th century, we find John Clare, in his poem *The Mores*, bemoaning the process of enclosure that produced the patchwork fields of Midland England. He was used to open fields, worked in common, where

> Cows went and came, with evening morn and night,
> To the wild pasture as their common right.

And complained that

> Inclosure came and trampled on the grave
> Of labour's right and left the poor a slave.

There is, therefore, a frequent identification of farming with tradition, even if what is seen as traditional might change over time. Farming, in this view, is seen as a way of life, in which doing right by the land, producing healthy crops and livestock, employing local people, and having a thriving and well-maintained farm to hand on to the next generation are more important than expansion, profit maximization, and integration with the food chain. There is none of the division between work and home or labour and leisure that characterizes the alienated urban worker. The whole family is involved in the farm, and they work with both mind and body, maintaining a close relationship with place and the local community. There is independence and dignity in what they do. Farmers have a sense of identity, emerging from their shared concerns, their reading of the agricultural press and membership of farmers' organizations, their shared leisure pursuits, even their style of dress and selection of friends and marriage partners.

Clearly this is an idealized image, but it is a long-established and powerful one. It is a version of the pastoral, that long-established literary form, written from the perspective of the town or court, in which the country is something other, and the simple rural virtues of honesty, peace, and innocence are contrasted with the courtly virtues of wit and cunning.

Perhaps more accurately it is the georgic, in that it idealizes a country life as one of satisfying work rather than idleness. In the US it has a long history as 'Jeffersonian Agrarianism', based on the idea that farmers are the most valuable citizens, with ideal social values. It emerges from the way in which people remember the countryside of their youth, but it is reinforced by images in books, in films (think of the image of the farmhouse in the film *Babe*), in photographs, and in paintings (Constable's *The Hay Wain* is perhaps the classic example). It also ignores the negative aspects of the image, the suspicion of strangers and the high rates of suicide and industrial accidents. In contrast to big, modern, mechanized, globalized agribusinesses it is sustainable, produces wildlife habitats and beautiful landscapes, and cares for the welfare of animals. How accurate this contrast might be is something that we shall examine in the following parts of the chapter.

Sustainability

The extent to which traditional farming in developed countries was sustainable is debatable. As we saw earlier, what counts as traditional depends not only on where you look, but when. The sort of farming that was carried out in North America and Europe in the middle of the 20th century, which present generations might now see as traditional, was already using fossil-fuel-powered tractors and fertilizers, although not to the extent to which they are used now. Tractors were beginning to make a significant impact from the 1930s, and from the middle of the 19th century some European farmers were using artificial fertilizers, especially phosphates.

The expansion of European animal production at the end of the 19th century was heavily reliant on cheap feedgrains imported from the US and Canada (which also supplied bread grains), so if we wish to go back to the point when Europe was feeding itself on entirely renewable resources we have to return to the middle of the 19th century, if not earlier. The population of Europe then was about 260 million, whereas today it is nearly three times as many.

In 1987, the World Commission on Environment and Development produced a report for the United Nations General Assembly, often called the Brundtland Report, from the Commission's chair, Gro Harlem Brundtland. It defined sustainability as meeting the needs of the present without compromising the ability of future generations to meet their own needs. Thus if the activities of our generation leave insufficient reserves of energy, water, or fertilizers for future generations, create global warming, pollute the environment, or destroy wildlife habitats and landscapes, they would be identified as unsustainable.

Clearly this has implications for agriculture. Different crops and animal enterprises have differing pollution effects, and, as we saw in the discussion of carbon footprints in Chapter 4, different crops and animal enterprises produce widely varying greenhouse gas emissions, implying that, from a climate change perspective, some are more sustainable than others.

But there is a further sustainability perspective too: will there be sufficient reserves of the finite resources that farming uses to cater for the needs of subsequent generations of farmers? This question applies to all kinds of inputs, from the soil and water that farmers have always used to the tractor fuel, electricity, fertilizers, and pesticides that are an integral part of modern farming in developed countries, and increasingly in developing countries, and it is controversial.

A critique of modern farming claims that a tonne of maize produced in the US for animal feed requires a barrel (42 US gallons or 159 litres) of oil, with two barrels of oil being needed to make the fertilizer and pesticides used on an average hectare of crops, so that agriculture is responsible for 7 per cent of the US's energy use. From a different perspective, another writer points out that agriculture accounts for only 3 per cent of global energy consumption, and argues that in times of scarcity oil tends to be reserved for vital uses, one of which would be food production.

Another important and finite resource is phosphorus. As we saw in Chapter 1, it is one of the three major constituents of the fertilizers that farmers use. Much of the total of 145 million tonnes that are mined each year goes into fertilizer, and at that rate the reserves that are worth exploiting at present prices and with present technology would, according to an estimate from the UN's Food and Agriculture Organization, be worked out in about 82 years. However, it is important to remember that the rate of use can vary considerably and that mining technology can change, so there are considerable disagreements about when the world will reach 'peak phosphorus'. Similarly with another principal fertilizer ingredient, potassium, for which the US Geological Survey calculates that the available reserves of 9.5 billion tonnes would suffice for 287 years at the present rate of consumption.

Nitrogen, the third major inorganic fertilizer constituent, is different. It is relatively easy to liquefy air and then distil it to separate out the nitrogen, but to be made into useful fertilizers such as ammonium nitrate or urea the nitrogen first needs to be combined with hydrogen to make ammonia. The usual feedstock for this process is natural gas, and making nitrogen fertilizers uses about 5 per cent of the annual global gas consumption.

Pesticides, being chemically much more complex and used in smaller total quantities, do not impose anything like the resource needs of fertilizers, but again, like fertilizers, have pollution implications that control the extent to which they can be used. One estimate of the cost of the damage caused by agriculture in the US to human health, water, soil, air quality, and wildlife suggested a range of figures between $US5.7 billion and $US16.9 billion in 2002, with a further $US3.7 billion in government expenditure on regulation and damage mitigation.

To be sustainable, therefore, either on a global or local scale, modern agriculture needs sufficient land of reasonable agricultural quality, with neither too much nor too little water on it, and a reliable supply of energy, fertilizers, and pesticides that can be used without producing major environmental problems. It is often argued that traditional farming methods are sustainable because they produce their own energy and fertilizers. The energy comes from the muscles of people or draught animals, both of which are fed by the food produced on the farm. The fertilizing nutrients are recycled in the farmyard manure produced by the animals (and often by the people too). Pests and diseases are controlled by selecting resistant varieties, mechanical weeding, fallowing, and crop rotation, or accepted as problems. Therefore, it is argued, this is the kind of farming that we must maintain where it still exists, or to which in the long run we must return. The contrary argument is that this kind of farming would produce too little food for the present world population. Already, according to one estimate, two-fifths of the world population are fed on food produced with the aid of nitrogen fertilizers.

Modern farming can, however, change to become more sustainable. One way of doing it is simply to adopt organic farming methods, in which pesticides and inorganic fertilizers are not used. Yields are lower, sometimes only half as much as conventional farming in the case of cereals, but very little lower in the case of dairy cows. Most organic farmers in developed countries still use tractors, so there is no saving in fossil fuel use. This form of farming usually accounts for only a small proportion of the land: in the United Kingdom, in 2004, for example, it was about 4 per cent of agricultural land.

But there are other approaches. One of the most dramatic responses to changing input availability has occurred in Cuba since 1990. In the 1980s, Cuban agriculture was dominated by large-scale sugar plantations, the produce of which was sold to the Soviet Union and other eastern bloc countries. It was heavily reliant on imported fertilizers, pesticides, oil, and cereals. With the break-up of the Soviet regime between 1989 and 1992, and a simultaneous trade embargo applied by the US, its main market disappeared and it could no longer afford its principal imports. Cuba's response was extraordinarily rapid. Within a few years the state farm sector was turned over to worker-run co-operatives operating on a much smaller scale, food crops and farm animals were increasingly found in the cities, farmers abandoned tractors for which there was little fuel and few spare parts and returned to animal traction, and synthetic fertilizers and pesticides were replaced by the use of compost and animal manures, resistant varieties, rotations, and intercropping, with predatory insects for pest control.

In a sense all this was simply a response to market forces: input prices rose and so did agricultural prices, so farmers had a greater incentive to produce food, but by using a different set of inputs. The government also played a part by changing the pattern of landholding, using Cuba's well-established network of agricultural scientists, and subsidizing food prices to consumers.

When insecticides first became available there was clearly a tendency to overuse them. In Sabah (Malaysia), in the 1960s, for example, the use of organochlorine insecticides on newly established cocoa crops affected both pest species and their natural predators, with the result that it was the pests that increased and destroyed the crop. After spraying was stopped, some of the pests were controlled using parasitic wasps and others by removing the trees that were their natural hosts.

There are numerous similar stories, and they have led farmers and agricultural scientists to realize that pests and pathogens can rapidly evolve resistance, that detailed study of their life histories can often suggest control methods, and that it is often better to live with low levels of pest attacks kept within bounds by a variety of control approaches than to attempt to wipe out the pest altogether.

In developed countries, too, increased fertilizer and pesticide prices, and changing agricultural policies, have encouraged farmers to find more sustainable farming methods. Instead of broadcasting granular fertilizer, some of which inevitably ends up in hedges and ditches, they spray liquid fertilizer, which can be placed more accurately. The imposition of Nitrate Vulnerable Zones in the UK requires farmers to spread fertilizer only at times of the year when crops are growing and so reduces losses to the water table. Those with smart phones can now download the Farm Crap App, which makes a visual assessment of manure and slurry application rates and calculates the nutrients provided, so producing potential savings in purchased artificial fertilizers.

Over 90 per cent of the UK cereal area is now sprayed with various herbicides, fungicides, and insecticides, and over half the area gets more than four spray treatments per year, but the total of active ingredients applied decreased from 15.5 million kg in 1990 to 9.4 million kg in 2010. These changes reflect the way in which society values what some might see as by-products of agriculture, in the form of pleasant landscapes and wildlife habitat, and the following section examines their relationship with farming in greater detail.

Wildlife and landscape

The American biologist Rachel Carson was one of the first writers to draw attention to the potential impact of modern agriculture on wildlife. She envisaged a time when pesticides would have wiped out the insects and weed species on which birds live, so that the birds themselves would no longer be there to sing. Hence the title of her book, *Silent Spring*, which was published in 1962. Since then, numerous other writers have drawn attention to the impact of modern farming methods on the environment.

In tropical countries the main concern has been over the loss of rainforest, with its impact on global warming. According to a report by the UN Food and Agriculture Organization in 2011, 7.1 million hectares (out of a total of 1.3 billion) were lost each year from the forests of the Amazon, the Congo Basin, and South East Asia in the 1990s (Figure 7). The annual figure decreased to 5.4 million hectares in the 2000s, more than a million of which were for the expansion of soya cultivation. Much of the rest, in South America, was the result of the expansion of cattle pasture.

7. Crop land displacing the Amazon rain forest.

In temperate countries too, agriculture affects wildlife and its habitats, and also the landscape. Clearly these two are linked, since the agricultural landscape is a wildlife habitat. But it is quite possible for farmers to create pleasant landscapes, with a patchwork of grass fields and waving corn, which offer little food or shelter to wildlife. Equally, an area of impenetrable overgrown scrub, which most people would consider to have little landscape value, might offer both food and habitat to a variety of mammals, birds, and insects. The following discussion will therefore make what might appear to be a rather artificial distinction between temperate agriculture's impact on wildlife and landscape.

The idea of wildlife and wilderness is, as Roderick Nash pointed out in his classic book *Wilderness and the American Mind*, dependent upon the existence of agriculture. Without fields in which plants grew because they had been sown there by humans, the term 'wilderness' was meaningless; and animals were just animals until the domesticated ones could be distinguished from the wild. Thus agriculture affected nature from its very beginning, and although modern agriculture has clearly had a major impact on wildlife, it is worth remembering that there is nothing new in this.

In 1566, for example, the English Queen Elizabeth I's government passed 'An Acte for the Preservation of Grayne', which remained in force until 1863 and provided for bounties to be paid for the destruction of what were deemed to be pest species of birds and mammals: a penny for the head of a bullfinch or the heads of twelve starlings or three rats, a halfpenny for the head of a mole, and so on. On the other hand, the activities of farmers also provided wildlife habitats, from the artificial woodland edges provided by hedges to the roosts for owls and bats in old barns.

In the last fifty years or so, however, the impact of modern agricultural techniques in temperate countries has reduced the extent and variety of wildlife habitat, the variety of wild plants, and the numbers of animals. Whereas the grass grew long in traditional late-cut hay meadows and sheltered partridges, corn buntings, and corncrakes, modern silage swards are cut shorter and earlier and consist of grass and few of the flowers that flourished in traditional meadows. Modern machinery works best in big fields, so many miles of hedge have been grubbed out, with a consequent loss of nesting sites and flowers. The impact of organochlorine pesticides on the breeding success of birds of prey in the 1950s and 1960s is well known. Their numbers consequently fell significantly, both in Western Europe and North America, where the peregrine became extinct east of the Rocky Mountains. More recently, the neonicotinoid insecticides, which only became widely used in the 1990s, have been identified as the cause of decreasing numbers of honeybees and also of insect-eating birds such as the skylark. These are just some of the better-known examples.

Farming has always affected the appearance of the landscape, from the time that the first farmers cleared land for their crops.

The chequerboard patterns of farms and square fields that the airline passenger sees while flying over the Midwestern states of the US were the result of government decisions made in the 19th century. The smaller scale farming landscapes of Western Europe evolved more gradually, with medieval open fields being converted at different times to smaller enclosed fields, which in recent years have again grown in size, while often new factory-like buildings have been erected to accompany the traditionally styled farmhouses and barns.

In addition, it is important to recognize that farming activities also change the landscape in the short term. Ploughed arable fields turn green as cereal crops germinate and then change to waving gold as harvest approaches. Up to the middle of the 20th century, cereal fields could also be red as a result of poppies growing among the corn, but the increased use of selective herbicides has now virtually eliminated that colour from the landscape. Great expanses of green oilseed rape (canola) become a vivid yellow when the crop flowers in the spring or early summer and then subside to a dull brown as the plants set seed. Even the colour of the cattle in the fields can change as a result of a farmer's decision: the brown, or brown and white, Shorthorns and Ayrshires that were found in dairying districts of Britain sixty years ago have now been almost entirely replaced by black and white Friesians and Holsteins.

Animal welfare

Animal welfare issues, like the questions of environmental sustainability discussed in the previous section, are easier to raise in developed economies where food is plentiful than in countries where food is scarce.

Keeping animals on farms implies that we accept the philosophical position that it is justifiable to keep other animals for the benefit of humans, and it is important to recognize that some people take the position that killing animals for human food is wrong, while others go further and argue that any form of animal exploitation, such as keeping them for their milk or wool, is unjustifiable. Clearly farmers who keep animals (and consumers who eat them and their products) do not take this position, but they would presumably all agree that abusing animals is wrong.

The more difficult question is to decide upon what constitutes abuse. Traditional small-scale farms, on which the farm staff are familiar with individual animals, should ensure the welfare of farm livestock; keeping large numbers of animals in close confinement is more likely to create physical and psychological deprivation and disease problems. This, at its simplest, is the way in which the potential animal welfare problem of modern farming is perceived, and stating it in this way suggests that there are different aspects of the problem: the extent to which the animal is prevented from exhibiting its normal behaviour; the effects of scale; and the question of disease.

Anyone who has seen animals who have been kept confined racing round the field into which they have been released before settling down to graze will recognize that they are more than biological machines. Seeing the apparent pleasure that a hen gets from a dust bath suggests that there might be more to her welfare than simply having enough to eat and drink. Conversely, animals kept on comfortable bedding and sheltered from cold, wind, and rain, or the blazing noonday sun in hotter climates, are presumably better off than those without such protection.

Traditional farming methods may still involve practices such as dehorning, castration, and branding which clearly interfere with the welfare of the animal, if only briefly, although practices such as debeaking chickens (i.e. cutting off the tips of their beaks so that they are less likely to peck each other) and docking the tails of pigs, both of which were designed to overcome intensive housing problems, are now much rarer. There are also degrees of confinement. Animals kept in groups in large pens are in a different position from a sow in a farrowing crate or hens in a battery cage, and advocates of housing point out that farrowing crates prevent sows from lying on their offspring, and that a stressed animal will be unproductive and prone to disease, so that no farmer will do anything deliberately to produce such stress.

This is where the second factor—the scale of operation—becomes significant. The level of expertise required to manage a large unit with hundreds or thousands of animals is of a different order of magnitude from that needed with traditional small flocks and herds, and it is much more difficult to give individual attention to each animal. As farmers have known for a long time, disease transmission is also easier when large numbers of animals are kept together. Hence old expressions of farming wisdom such as 'a sheep's worst enemy is another sheep'.

High productivity can also result in extra stress on an animal. A cow that is producing a lot of milk needs a big udder, which imposes extra strain on the attachment ligaments, and she will also be more prone to mastitis (an udder infection) and foot problems. Liver abscesses have been found in feedlot cattle fed on highly concentrated, low roughage diets. All these are problems for farmers, but they are also questions that need to be confronted by society in general when reflecting upon the kind of farming that we want to produce our food.

Chapter 6
Farming futures

Little so far has appeared in this book about government involvement in agriculture, but national governments as well as international organizations—such as the Food and Agriculture Organization of the United Nations and the World Trade Organization—all have an interest in agriculture. So too do numerous non-governmental organizations, pressure groups, and charities.

In this chapter we shall look at some of the most important questions that agricultural policymakers currently have to think about: what will be the impact of climate change; what is the impact of genetically modified organisms (GMOs) and are they safe to use; and will we be able to feed the world in the future? But first we need to look at some of the arguments around agricultural policy.

Governments and farming

Most governments in industrialized countries support their agricultural industries in one way or another, but the extent to which they do so varies considerably. Whereas farmers in Australia and New Zealand get hardly anything, those in Norway, Iceland, Switzerland, Japan, and South Korea received between one-half and two-thirds of their receipts from government subsidies in 2011–13. Since in most cases agriculture forms only a small part of the economy of industrialized countries we might wonder why they support farmers, and why they often have government ministries with specific responsibility for agriculture, but not, for example, for retailing, which employs more people and contributes more to most developed economies. And since, in capitalist economies, markets are supposed to ensure the fulfilment of consumers' wants, why would governments in such societies want to intervene in agricultural markets at all?

Logically, the answer must be that agricultural markets do not produce what societies and governments want. They fail to do so in several different ways. First, they are not good at dealing with price fluctuations. The demand for agricultural products is fairly constant, because food consumers want to eat every day, and in roughly similar amounts, except perhaps at festivals such as Christmas. The supply of these products, however, can change considerably as a result of weather and disease changes. The result, as we saw in Chapter 3, is a fluctuation in price. High prices make life difficult for poorer consumers, who have to spend a high proportion of their incomes on food. Low prices may mean that marginal farms, which might survive in the long run at average price levels, are driven out of business by one or two low-income years. Consequently there has in the past been widespread agreement that dealing with such price fluctuations is worthwhile. The problem is that policies to control price fluctuations may become policies to raise farm prices and/or incomes, and these are more controversial.

This takes us on to other ways in which agricultural markets fail. One problem is that the firms to which farmers sell normally have much more market power—power to influence prices through their own decisions—than farmers, simply because they are bigger. Another is that the market does not normally pay for the ecosystem services, in the form of wildlife habitats and pleasant landscapes, that farmers deliver as a by-product of food production. Yet another is that the demand, in industrialized countries, for the basic food commodities that farmers produce does not increase very rapidly (see Chapter 3). Consequently, if farmers adopt new technologies and increase output, prices will tend to fall. In addition, they may face competition from farmers in other countries who can produce profitably at lower prices.

Farm lobby groups argue that governments in industrialized countries should tackle these issues. Their argument is that a country should be able to feed itself, or at least maintain a reasonable level of self-sufficiency, for several reasons.

One is national security. Governments have a responsibility to see that food is available in an emergency, although it might be argued that stable trading relationships are just as effective a form of food security, because food exporters are as concerned to ensure access to markets as importers are to secure supplies. Another argument concerns employment: a run-down agricultural sector means not only jobs lost in the countryside but also declining employment in the industries that supply farmers, such as fertilizers and farm machinery, and in the food industry that processes their products. There is also the question of paying for imported food. Governments have to decide whether their national resources are best employed in producing their own food or in producing something that can be sold in return for food.

Even if, on weighing up all these considerations, a government decides against intervening in agricultural markets, farmers might still argue that they are faced with declining and fluctuating incomes through no fault of their own. In the UK, the income per farmer in 2000 was only about a quarter of its 1976 level, although by 2010 it had recovered to just over half of that level. Since other groups in society (old-age pensioners and the unemployed, for example) receive state support in these circumstances, shouldn't farmers receive it too? On the other hand, there are different social groups, such as small shopkeepers and steelworkers, who work in traditional or declining industries and receive little or no state assistance, and who, unlike owner-occupying farmers, do not have a large capital asset such as a farm to fall back on.

These ideas should make it clear that questions of farm income-support involve value judgements which societies normally resolve through political mechanisms. Thus the function of political pressure groups, such as the National Farmers' Union (NFU) in the UK, or the American Farm Bureau Federation, the National Grange, and the National Farmers Union in the US, or the Comité des Organisations Professionnelles Agricoles (COPA) in the EU, is to put the farmers' case. Other groups, such as those representing the food industry, or environmentalists, may argue differently.

The history of agricultural policy in industrialized countries suggests that governments respond both to effective lobbying and to changing economic and political circumstances. In the US, intervention in agricultural markets can be traced back to the Agricultural Adjustment Acts of the 1930s, which not only allowed the federal authorities to buy grain from farmers, but also to pay them for not growing food on part of their land. This was the policy ridiculed by Joseph Heller in his novel *Catch-22*, in which Major Major's father was paid for not growing alfalfa, so he didn't grow more, and with the money the government paid him he bought more land upon which he didn't grow still more.

US agricultural policy is still concerned with evening out market fluctuations. Among the provisions of the Agricultural Act of 2014 farmers can be paid various sums from the public purse when market prices are low, or, in the case of dairy farmers, when the gap between milk prices and feed costs falls below a certain trigger level, at which point the Secretary of Agriculture can instigate purchases of dairy products to be used in domestic food programmes.

Successive UK governments in the late 19th and early 20th centuries relied on imports of agricultural products, especially from the Americas and Oceania, until the late 1930s. Then sea-borne trade, which at that time accounted for nearly two-thirds of UK food supplies, came under threat from surface warships and submarines. In response, the government initiated a series of price subsidies to promote increased domestic agricultural output. These were maintained after World War II because the UK economy was no longer in a position to pay for the previous level of imports, and were successful in raising the domestic self-sufficiency level to roughly two-thirds of domestic consumption by 1973.

The UK then joined what became the EU, which had a Common Agricultural Policy (CAP) operating on a different basis. It maintained domestic prices at higher than world levels using import tariffs, while surplus produce was initially bought into official stores and subsequently sold to the world market with the aid of export subsidies. This policy was very successful in doing what it was initially designed to do, which was to increase European agricultural output. But by the 1980s it was attracting considerable criticism as a result of its cost, its environmental impact, and its disruption of world agricultural markets. In consequence, it has undergone radical change.

Barriers to importing into the EU, which is the world's largest agricultural importer by a big margin, have been reduced, and farmers' incomes are now supported by direct payments. These do not encourage them to produce more than can be sold on domestic and export markets, but they are dependent upon the size of the farm business, so that big farmers receive more money than small farmers. This leads, in some cases, to very rich people receiving money from the public purse; whether this is justified by the argument that it represents a payment for the environmental services they provide is for the reader to decide.

In developing countries and emerging economies the variety of agricultural policies is as great, if not greater, than in industrialized countries, and has included measures to raise or stabilize prices, input subsidies, investment in education and technical training, research and development, advisory work, development of credit institutions and producer co-operatives, and land reform.

Part of the problem is that the opinions of experts and politicians on the most effective way of promoting development have changed over time. Fifty years or so ago it was thought that agricultural development was a prerequisite for development of the economy as a whole. Then it appeared that investment in education and health paid off more quickly, so in several countries farm prices were kept low and investment concentrated in urban areas. More recently opinion has swung back towards promotion of agriculture. At the same time, opinions on the best ways of doing so have changed. Whereas in the 1960s the emphasis was on research and development and subsidies for seeds and fertilizers, it later changed to attempts to identify the most appropriate price levels, but recently renewed attention has been given to R&D and the World Bank has advocated seed and fertilizer subsidies.

Climate change

The reality of climate change is now widely accepted, although a survey of 1,276 farmers in Iowa in 2011 found that 32 per cent of them thought that climate change was not occurring, or that there was insufficient evidence of it. The International Panel on Climate Change (IPCC) has predicted a global average temperature rise of between 2°C and 4°C above that of the pre-industrial climate by 2050. To put it another way, global average temperatures rose by about 0.13°C per decade between 1950 and 2010 and are projected to rise by about 0.2°C per decade in the next three decades.

These increased temperatures are likely to increase the level of activity in the hydrological cycle, which in simple terms implies more rain. Where it will fall and when is more difficult to predict, especially in monsoon areas, although changes in the timing and pattern of seasonal rainfall seem likely. Higher temperatures also affect polar ice caps and mountain snow melt, leading to a predicted rise in sea level of two metres by 2100. It is also possible for climatic change in one region to have impacts elsewhere, the most obvious example being that less snow in the Himalayas implies more drought on the plains around the Indus and Ganges rivers. In addition to these gradual changes, there are also likely to be increases in climatic variability and extreme weather events, meaning longer and more frequent droughts, an increased risk of extreme wet weather leading to floods in the northern hemisphere, and fewer but longer and more intense tropical cyclones.

Agriculture

Agriculture and climate change have a twofold relationship. Climate has an impact on agriculture, and agriculture has an impact on climate. Taking the latter first, we have already discussed (in Chapter 4) agriculture's carbon footprint, pointing out that agriculture makes a significant contribution to the world's greenhouse gas emissions, even if its precise extent is disputed. A high proportion—one estimate is 75 per cent—of these emissions originate in low- and middle-income countries, where rice production and biomass burning are concentrated. Consequently farmers are under pressure to change, both in response to policymakers and in their own self-interest. They may need to upgrade their drainage systems to cope with increased rainfall, or adapt their cropping to shorter growing seasons or more frequent droughts. In order to reduce greenhouse gas emissions they could increase soil carbon by changing the crops they grow or the cultivation methods they use. They could also process animal and crop wastes in anaerobic digesters to collect methane instead of allowing it to escape to the atmosphere, and increase the use of organic manures in order to reduce their fertilizer consumption.

As far as the impact on agriculture is concerned, perhaps the most dramatic of the expected effects of global warming, apart from catastrophic storms and floods, is the loss of agricultural land as a result of rising sea levels. In coastal zones, and especially in the river deltas of Bangladesh, Myanmar, India, Pakistan, and Egypt, one estimate suggests that some 650 million people could be affected. Conversely, reduced snowfall in the Himalayas is predicted to produce water shortages in the dry season many miles away in the basins of the Indus, Ganges, and Brahmaputra rivers, relied upon by 500 million people for domestic and agricultural purposes. In Africa, the amount of water in the Niger river basin decreased in the 20th century, and the drought in Russia in 2010 reduced wheat yields by 40 per cent.

Increased temperatures and changing rainfall patterns can restrict the length of the growing season. In northern Ghana, for example, farmers remember the rainy season being long enough to get two harvests, whereas now the second crop is at risk if the rains fail. Over the world as a whole, about 80 per cent of the agricultural area is rain-fed (the proportion is higher in Africa and Latin America), but the remaining 20 per cent of irrigated land produces over 40 per cent of the world's food, and is thought to be vulnerable to climate-change-induced water shortages.

Temperature fluctuations also affect crop yields, especially high temperatures (above 32°C) during the flowering stage of cereals. Above this temperature rice yields have been shown experimentally to be reduced by 90 per cent. Every degree-day above 30°C in maize (corn) crop trials reduced the final yield by 1 per cent under optimal rainfall conditions and 1.7 per cent in a drought.

However, it is important to remember that higher temperatures are likely to be beneficial in temperate regions, since the potential crop-producing area could increase, as will the length of the growing season, and possibly rainfall, while cold periods decrease.

Higher carbon dioxide levels in the atmosphere also have a fertilizing effect on crop growth (see Chapter 1). Consequently agriculture in North America, southern Latin America, and Europe is less likely to be adversely affected than sub-Saharan Africa and Asia, and could see yield increases in some crops.

Conversely, warmer temperatures also affect the ability of pests to survive and thrive. Aphids and weevil larvae have been shown to grow faster in higher carbon dioxide concentrations, and warmer winter temperatures would reduce seasonal mortality, allowing earlier and greater dispersion of pest species. Animal health is also likely to be affected. Perhaps the best-known recent example is in the spread of blue tongue disease in cattle and sheep to north and west Europe. The blue tongue virus is transmitted by the bite of midges of the *Culicoides* genus, and has been recognized for a long time around the Mediterranean, with *C.imicola* as the insect vector. It appears that the virus has recently been acquired by other midge species such as *C.obsoletus* and *C.pulicaris*, and their ability, and that of the virus, to survive has been increased by milder winters. Hence the spread of the disease, which reached south-east England by 2007.

Given the wide range of possible effects, it is not easy to predict the overall impact of climate change. A study of temperature and precipitation effects between 1980 and 2008 concluded that temperature had had a bigger impact than precipitation, and that maize production was 3.8 per cent lower, and wheat 5.5 per cent lower, than either would have been in the absence of the climate trends in that period. Gains and losses in rice and soya bean production roughly evened out, with rice output in temperate areas benefiting from higher temperatures. There could be big differences between countries: wheat yields in Russia were reduced by almost 15 per cent, whereas there were no effects in the US because there were no significant climate trends there. The authors of the study admitted that their model was likely to be pessimistic because it neglected the effects of carbon dioxide fertilization. Taking these into account, they estimated the total increase in average commodity prices resulting from climate change over these twenty-eight years at 6.4 per cent.

Using studies like this in attempting to predict the future is, however, fraught with difficulty. Not only do they neglect the possibility of farmers adapting to changing circumstances, they also neglect changes in the wider world within which farmers will be living and working. Hence the IPCC's idea of a Climate Vulnerability Index, which combines seventeen different variables, not only climatic but also demographic, socio-economic, and measures of governance. This clearly identifies much of Africa and Asia as most vulnerable, and North America and most of Europe as least vulnerable.

GMOs

Traditional breeders worked by attempting to combine the desirable traits of different varieties or breeds of the same species. Early in the 20th century, for example, Professor Biffen at Cambridge crossed an established English wheat variety called Squareheads Master with a Russian variety, Ghirka, which was resistant to yellow rust, a fungal disease of the foliage of the wheat plant, to produce a new rust-resistant variety called Little Joss. Similarly, animal breeders have crossed a sheep breed with a high milk yield with another breed with a high lambing percentage in order to produce breeding ewes capable of producing plenty of milk for twin lambs. Hence the importance of rare breeds of livestock and exotic plant varieties. They represent a reservoir of genes with different properties that are available for future use in response to unpredictable changes in requirements.

The problem with these traditional approaches is that they are time consuming, so it is expensive to produce new breeds or varieties. Work on producing a potato variety resistant to attack by the fungus causing blight (see Chapter 1) began at the Scottish Plant Breeding Station (now part of the James Hutton Institute) in 1937. The variety Pentland Ace, which incorporated genes from a blight-resistant Mexican potato, was not released until 1951, and this is said to be an example of a relatively rapid breeding programme. Moreover, there are limits to these conventional methods: if a trait (e.g. blight resistance) is not present somewhere in the genes of a species it cannot be invented. From the 1920s crop breeders tried to increase genetic variation by artificially inducing mutations using, for example, X-rays or gamma rays. This is a random technique requiring very large numbers of plants to produce a few useful individuals, but it has been used to produce new varieties of oilseed rape (canola), rice, and barley.

Genetic modification offers, in theory at least, a more rapid and precise way of incorporating new traits into crops. Several stages are involved. First an organism must be found with the desired characteristics. The second stage is the isolation of the gene responsible for producing the desirable trait in the donor organism, which is done by using restriction enzymes to cut its DNA at the desired location. This gene is then cloned—copied many times—before being inserted into the target plant using a 'gene gun' or a bacterium. Using a gene gun means attaching the cloned genes to tiny metallic particles which are then inserted into the target plant cells under high pressure so that the donor DNA is integrated into the DNA of the target plant inside its cell nuclei. It sounds rather crude, but it has been used successfully, especially for monocotyledon crops such as wheat or maize.

For dicotyledonous plants such as potatoes, tomatoes, or carrots, the preferred technique is to use a bacterium, *Agrobacterium tumifaciens*, a natural plant parasite, to transport the donor DNA into the target plant. It's then a matter of testing the resulting plants to check that they have the desired characteristics, multiplying them, and putting them on the market.

One of the early, and still important, uses of the technique was to produce a soya bean variety that was resistant to glyphosate, a herbicide that kills nearly all green plants and is sold under the trade name RoundUp. Researchers working for Monsanto, the firm which produced glyphosate, found a bacterium that was resistant to it, and incorporated the gene responsible into a soya bean variety that they named RoundUp Ready. By the mid-1990s it was on the market, enabling farmers to control weeds by spraying with glyphosate without affecting the crop (and also increasing glyphosate sales).

Another well-known example was the ability of the soil bacterium *Bacillus thuringiensis* (Bt) to produce a protein toxic to insect larvae by affecting the cell membranes in their guts. Different strains of the bacterium produce varieties of the protein that affect different types of insects. Again the relevant genes could be identified. They were incorporated into several maize and cotton varieties by the mid-1990s and provided protection against the European corn borer, a major pest of maize, and the cotton bollworm and pink bollworm, major pests of cotton. Subsequent developments have resulted in the 'stacking' of traits, meaning that there are now maize varieties that have both Bt and glyphosate-resistant genes.

Genetic modification has also been applied, with varying degrees of success, to producing resistance to drought and to virus attacks. Rice varieties (so-called 'golden rice') have been engineered to have an increased vitamin A content and so help to protect against blindness. There are soya bean varieties with higher concentrations of omega-3 fatty acids, which are thought to help reduce cardiovascular disease. More experimentally, it has been used to produce transgenic farm animals. For example, genes from spinach and nematodes have been incorporated into pigs and dairy cattle to increase the level of omega-3 fatty acids in meat and milk, in order to reduce the risks of cardiovascular disease in people. So far, however, GM farm animals have remained within the laboratory.

In the twenty years or so since they were first introduced, the area sown to GM crops has increased enormously. A GM advocate claims that it has increased from 1.7 million hectares in 1996 to 170 million hectares across the world in 2012, evenly split between industrialized and developing countries. Much of this area is in the four crops that were involved from the beginning: soya beans, for which three-quarters of world production is now from GM varieties, cotton (nearly half GM), maize, and oilseed rape (in each case over a quarter GM). In some countries, such as the US, the proportion is higher, with most of the soya beans and rapeseed, and much of the cotton and maize, now coming from GM varieties.

There has also been an expansion in the range of crops involved, with GM versions of alfalfa, apples, courgettes, sugar beet, sugar cane, and wheat now available. Most of the Hawaiian papaya crop now comes from a variety that has had a gene added to produce resistance to the papaya ringspot virus.

Nevertheless, GM crops remain controversial, with most European countries banning, or carefully controlling, their use. Opponents of GM argue that combining modern scientific techniques such as genome sequencing and gene marker mapping, coupled with conventional breeding methods, are as useful as genetic engineering as far as increasing yields are concerned.

There is insufficient space in this book to go into all the arguments about GM in detail; listing them will have to do. There are fears that some people may develop allergies to GM varieties or find them toxic. Perhaps more significant are worries that modified genes might spread unintentionally, with unpredictable effects on non-target organisms. A study of GM maize being grown in Germany found that it was difficult to prevent wind-blown GM maize pollen from fertilizing non-GM maize plants. Similar studies in South Africa and Ghana led to concerns about the maintenance of traditional maize varieties.

It is also important to remember that plants and animals naturally vary in their susceptibility to toxicity, so that killing the most susceptible advantages those least so, and thus resistance will tend to build up in plant and animal species. Reuters reported in September 2011 that twenty-one glyphosate-resistant weed species, including marestail in Ohio and waterhemp in north-east Kansas, affected 11 million acres in the US. In order to control the development of resistance among pest species, farmers in the US are required to plant part of their crop areas with non-Bt varieties to act as refuges for non-resistant insects, but policing this policy would probably be difficult in other parts of the world. The Monsanto website blamed the breakdown of pink bollworm resistance in one of its Bt cotton varieties in India on the failure of farmers there to plant non-Bt refuge crops.

There are also social and ethical concerns. Since GM varieties normally have to be purchased afresh from the seed companies each year, the traditional practice of saving seed is no longer available, which can affect poor farmers in particular while benefiting the multinational companies that produce them. The effect that these actual or potential problems will have on the current expansion of GM crops remains to be seen.

Feeding nine billion people

In 2010 the UN Population Division's medium estimate predicted that there would be 9.3 billion people in the world in 2050. In 2013 a *New York Times* article calculated that agriculture currently produces 2,700 calories for each of the 7 billion inhabitants of the world. If there were to be no increase in food output, current production would still produce just over 2,000 calories per person for the 2050 population, which is perfectly adequate for survival. So is there anything to worry about?

If you've read this far you will have realized that it's not quite as simple as that. About a third of those calories currently go to feed animals, some will be wasted as they go through the food chain, and at present about 5 per cent will be converted into biofuels. The remaining supplies are not evenly distributed, because the industrialized countries get more per person than developing countries, and richer people everywhere can spend more on food than poorer people. Not only that, but people in the future may have different food demands, and the resources and technologies available to produce food will be different. This book, let alone the remains of this chapter, is far too short to produce a considered estimate of the likely world food balance in 2050, but we can at least look at some of the things to think about in doing so.

As we saw in Chapter 3, the two main factors affecting future food demand are the number of people in the world and the amount of money they have to spend. That medium population estimate of 9.3 billion assumes that the average woman will have 2.17 children in 2045–50. In 2005–10 she had 2.52 children. But if she has 2.64 children, the high estimate, there could be 10.6 billion mouths to feed. Factors affecting female fertility, such as education, income, access to birth control, religion, and culture are therefore crucial as far as future food demand is concerned. On the whole, as societies become richer, people tend to have fewer children. But they also eat more, and have a more varied diet, as we also saw in Chapter 3. We have already seen an increase in demand for pork, and consequently in demand for feed grains, as the Chinese economy has grown, and we would expect other emerging economies to increase their demand for animal protein and fats as their consumers spend their increased incomes. In short, we should expect the demand for food to grow by more than the growth in population, but by how much more is considerably more difficult to predict.

What we do know is that in the past increasing demand, and especially increasing population, has been more than met by increases in supply. The historian Giovanni Federico calculated that while the world population rose six or sevenfold between

1800 and 2000, world agricultural production rose by at least ten times in the same period. In the 19th century most of the increase came from land in North and South America, and Australia and New Zealand, that had not previously produced food and fibre for the world market. In the 20th century the increase was mainly the result of increases in output from each hectare of land and/or person working on it. Can either of these solutions be applied again in the 21st century?

We are still cutting down trees in countries such as Argentina and Brazil in order to grow maize and soya beans, but this is increasingly coming under political pressure, and, as we saw earlier, we are likely to lose land as a result of sea-level rises, droughts, and desertification as global warming continues. It is not easy to predict how these trends will balance out in detail, but it does seem clear that we cannot expect anything like the sort of extra-land-based production increase that we saw in the 19th century.

That leaves us with producing more per hectare of agricultural land. In the past, faced with an increase of demand relative to supply, and consequently increased prices, farmers have been good at producing more by using extra inputs—from more hand labour in weeding to extra fertilizers and pesticides. Scientists and technologists have also been good at producing better plant varieties, livestock breeds, machinery, and so on. While world population increased by just over 80 per cent between 1970 and 2009, world maize production increased three times and wheat and rice production more than doubled. Can this continue into the future?

In industrialized countries, the scope for easy ways to increase output may be limited, but there already exists a highly trained and sophisticated agricultural labour force that is used to adopting new technologies such as integrated pest management in their search for the desirable realization of sustainable intensification.

The greatest opportunities for increasing output per hectare are also where the increase in demand is likely to be greatest: in the emerging and developing economies. They are the ones in which a little extra fertilizer, or slightly better seeds, or a little extra draught power from cattle or donkeys, can make the greatest difference. Farmers can use mobile phones both to receive technical advice and also to find out which local markets have the best prices. Even simple things such as a nearby piped water supply, or a way of cooking food that does not require a woman to spend two or three hours a day searching for firewood, could mean that more time is available for weeding a crop or growing a second crop or looking after animals. Bringing the world average yields of crops and livestock up to the levels already being achieved on the more productive farms is therefore the challenge for the 21st century.

Further reading

General

N. De Pulford and J. Hitchens, *Behind the Hedge: Where Our Food Comes From*, Ammanford: Sigma Press, 2012.

C. Martiin, *The World of Agricultural Economics: An Introduction*, London: Routledge, 2013.

E. Millstone and T. Lang, *The Atlas of Food: Who Eats What, Where, and Why*, Berkeley: University of California Press, revised edn., 2008.

R. J. Soffe (ed.), *The Agricultural Notebook*, 20th edn, Oxford: Blackwell, 2003.

C. C. Webster and P. N. Wilson, *Agriculture in the Tropics*, 2nd edn, London: Longman, 1980.

On the history of agriculture

D. B. Danbom, *Born in the Country: A History of Rural America*, 2nd edn, Baltimore, MD: Johns Hopkins University Press, 2006.

M. B. Tauger, *Agriculture in World History*, London: Routledge, 2011.

J. Thirsk, *Alternative Agriculture: A History, from the Black Death to the Present Day*, Oxford: Oxford University Press, 1997.

Chapter 1: Soils and crops

On soils

R. J. Buresh, P. A. Sanchez, and F. Calhoun (eds), *Replenishing Soil Fertility in Africa*, Madison, WI: Soil Science Society of America, 1997.

E. A. FitzPatrick, *An Introduction to Soil Science*, 2nd edn, Harlow: Longman, 1986.

T. R. Paton, G. S. Humphreys, and P. B. Mitchell, *Soils: A New Global View*, London: UCL Press, 1995.

C. Reij, I. Scoones, and C. Toulmin (eds), *Sustaining the Soil: Indigenous Soil and Water Conservation in Africa*, London: Earthscan, 1996.

E. J. Russell, *Russell's Soil Conditions and Plant Growth* (ed. A. Wild), 11th edn, Harlow: Longman, 1988.

On crops

J. Christopher, *The Death of Grass*, London: Michael Joseph, 1956.

H. J. S. Finch, A. M. Samuel, and G. P. F. Lane, *Lockhart and Wiseman's Crop Husbandry*, 8th edn, Cambridge: Woodhead Publishing, 2002.

S. Meakin, *Crops for Industry: A Practical Guide to Non-Food and Oilseed Agriculture*, Marlborough: Crowood Press, 2007.

J. W. Purseglove, *Tropical Crops: Dicotyledons*, vols 1 and 2, London: Longmans, 1968.

J. W. Purseglove, *Tropical Crops: Monocotyledons*, London: Longmans, 1972.

N. W. Simmonds and J. Smartt, *Principles of Crop Improvement*, 2nd edn, Oxford: Blackwell, 1999.

W. D. Walters, *The Heart of the Cornbelt: An Illustrated History of Corn Farming in McLean County*, Bloomington, IL: McLean County Historical Society, 1997.

Chapter 2: Farm animals

T. G. Field and R. E. Taylor, *Scientific Farm Animal Production: An Introduction to Animal Science*, 9th edn, Upper Saddle River, NJ: Prentice Hall, 2008.

R. D. Frandson and T. L. Spurgeon, *Anatomy and Physiology of Farm Animals*, Philadelphia, PA: Lea and Febiger, 1992.

P. McDonald, R. A. Edwards, J. F. D. Greenhalgh, C. A. Morgan, L. A. Sinclair, and R. G. Wilkinson, *Animal Nutrition*, 7th edn, Harlow: Pearson Education, 2011.

M. L. Ryder, *Sheep and Man*, London: Duckworth, 1983.

R. T. Wilson, *The Camel*, Longman, 1984.

Chapter 3: Agricultural products and trade

G. L. Cramer, C. W. Jensen, and D. D. Southgate, *Agricultural Economics and Agribusiness*, 8th edn, New York: Wiley, 2001.

O. Ecker and M. Qaim, 'Income and Price Elasticities of Food Demand and Nutrient Consumption in Malawi'<http://ageconsearch.umn.edu/bitstream/6349/2/451037.pdf>.

A. Regmi, M. S. Deepak, J. L. Seale Jr, and Jason Bernstein, 'Cross-Country Analysis of Food Consumption Patterns' <http://www.ers.usda.gov/media/293593/wrs011d_1_.pdf> (USDA elasticities data).

R. Tiffin, K. Balcombe, M. Salois, and A. Kehlbacher, 'Estimating Food and Drink Elasticities' <https://www.gov.uk/government/uploads/system/uploads/attachment_data/file/137726/defra-stats-food-farm-food-price-elasticities-120208.pdf>.

WTO, 'Merchandise Trade by Product' <http://www.wto.org/english/res_e/statis_e/its2013_e/its13_merch_trade_product_e.pdf>.

The website of the UN Food and Agriculture Organization, <http://faostat.fao.org/>, is an excellent source of global agricultural statistics.

Chapter 4: Inputs into agriculture

D. Blandford, I. Gaasland, and E. Vårdal, 'Extensification versus Intensification in Reducing Greenhouse Gas Emissions in Agriculture: Insights from Norway', *Eurochoices* 12(3), 2013, pp. 4–8.

FAO, *FAO Statistical Yearbook 2013: World Food and Agriculture* <http://www.fao.org/docrep/018/i3107e/i3107e00.htm>.

S. K. Lowder, J. Skoet, and S. Singh, 'What Do We Really Know about the Number and Distribution of Farms and Family Farms Worldwide?', Background paper for *The State of Food and Agriculture 2014*, ESA Working Paper No. 14–02, Rome: FAO, 2014.

M. M. Mekonnen and A. Y. Hoekstra, 'The Green, Blue and Grey Water Footprint of Crops and Derived Crop Products', *Hydrology and Earth System Sciences* 15, 2011, pp. 1577–600.

M. M. Mekonnen and A. Y. Hoekstra, 'A Global Assessment of the Water Footprint of Farm Animal Products', *Ecosystems* 15, 2012, pp. 401–15.

P. Scarborough, P. N. Appleby, A. Mizdrak, A. D. M. Briggs,
R. C. Travis, K. E. Bradbury, and T. J. Key, 'Dietary Greenhouse
Gas Emissions of Meat-Eaters, Fish-Eaters, Vegetarians and
Vegans in the UK', *Climate Change* 125, 2014, pp. 179–92.

Chapter 5: Modern and traditional farming

P. Brassley, 'Agricultural Technology and the Ephemeral Landscape',
in D. E. Nye (ed.), *Technologies of Landscape: From Reaping to
Recycling*, Amherst: University of Massachusetts Press, 1999,
pp. 21–39.

D. Briggs and F. Courtney, *Agriculture and Environment: The
Physical Geography of Temperate Agricultural Systems*, Harlow:
Longman, 1989.

J. Burchardt and P. Conford, *The Contested Countryside: Rural
Politics and Land Controversy in Modern Britain*, London:
I.B.Tauris, 2008.

G. Conway, 'The Doubly Green Revolution', in Pretty (2005)
pp. 115–27.

FAO and International Tropical Timber Organisation, *The State of
Forests in the Amazon Basin, Congo Basin, and South-East Asia*,
Rome: FAO, 2011.

R. Gasson and A. Errington, *The Farm Family Business*, Wallingford:
CAB International, 1993.

C. A. Hallmann et al., 'Declines in Insectivorous Birds are Associated
with High Neonicotinoid Concentrations', *Nature* 511, 17 July 2014,
pp. 341–3.

R. Lovegrove, *Silent Fields: The Long Decline of a Nation's Wildlife*,
Oxford: Oxford University Press, 2007.

P. Lymbery and I. Oakeshott, *Farmageddon: The True Cost of Cheap
Meat*, London: Bloomsbury, 2014.

P. McMahon, *Feeding Frenzy: The New Politics of Food*, London:
Profile Books, 2013.

R. Nash, *Wilderness and the American Mind*, 3rd edn, New Haven,
CT: Yale University Press, 1982.

J. Pretty (ed.), *The Earthscan Reader in Sustainable Agriculture*,
London: Earthscan, 2005.

P. Rosset and M. Bourque, 'Lessons of Cuban Resistance', in Pretty
(2005) pp. 362–8.

V. Smil, *Enriching the Earth: Fritz Haber, Carl Bosch, and the Transformation of World Food Production*, Cambridge, MA: MIT Press, 2001.

E. M. Tegtmeier and M. D. Duffy, 'External Costs of Agricultural Production in the United States', in Pretty (2005) pp. 64–89.

W. D. Walters, *The Heart of the Cornbelt: An Illustrated History of Corn Farming in McLean County*, Bloomington, IL: McLean County Historical Society, 1997.

R. Williams, *The Country and the City*, London: Chatto and Windus, 1973.

UK Pesticide use statistics at <http://pusstats.fera.defra.gov.uk>.

For phosphate reserves: <http://www.fao.org/docrep/007/y5053e/y5053e07.htm> (consulted 19.1.2015).

For potassium reserves: <http://minerals.usgs.gov/minerals/pubs/commodity/potash/mcs-2011-potas.pdf> (consulted 19.1.2015).

Chapter 6: Farming futures

On agricultural policy

J. Brooks, 'Agricultural Policy Choices in Developing Countries' <http://economics.ucr.edu/seminars_colloquia/2010/applied_economics/Brooks%20paper%20for%201%204%2010.pdf>.

P. de Castro, 'A Comparative Approach to European and American Agricultural Policies', *momagri.org* <http://www.momagri.org/UK/points-of-view/A-comparative-approach-to-European-and-American-agricultural-policies_798.html>.

European Commission, 'The Common Agricultural Policy after 2013' <http://ec.europa.eu/agriculture/cap-post-2013/index_en.htm>.

European Commission CAP overview <http://ec.europa.eu/agriculture/cap-overview/2014_en.pdf>.

European Commission, *Agriculture in the European Union Statistical and Economic Information: Report 2013* <http://ec.europa.eu/agriculture/statistics/agricultural/2013/pdf/overview_en.pdf>.

P. Shelton, 'Can the US Farm Bill and EU Common Agricultural Policy Address 21st Century Global Food Security?', *IFPRI Blog*, 23 July 2014<http://www.ifpri.org/blog/can-us-farm-bill-and-eu-common-agricultural-policy-address-21st-century-global-food-security>.

On climate change

Anon., *Climate Impacts on Food Security and Nutrition: A Review of Existing Knowledge*, Exeter: Met Office and UN World Food Programme, 2012.

D. B. Lobell, W. Schlenker, and J. Costa-Roberts, 'Climate Trends and Global Crop Production Since 1980', *Science* 333, 29 July 2011, pp. 616–20.

L. Nijs, *The Handbook of Global Agricultural Markets: The Business and Finance of Land, Water and Soft Commodities*, Basingstoke: Palgrave Macmillan, 2014, chapter 3.

C. L. Walthall et al., *Climate Change and Agriculture in the United States: Effects and Adaptation*, Washington, DC: USDA Technical Bulletin 1935, 2013, available at <http://www.usda.gov/oce/climate_change/effects.htm>.

On GM crops

R. Boyle, 'How to Genetically Modify a Seed, Step By Step', *Popular Science* <http://www.popsci.com/science/article/2011-01/life-cycle-genetically-modified-seed?single-page-view=true>.

F. Forabosco et al., 'Genetically Modified Farm Animals and Fish in Agriculture: A Review', *Livestock Science* 153 (May 2013) pp. 1–9.

N. Halford, *Plant Biotechnology*, London: Wiley, 2006.

C. James, 'Global Status of Commercialized Biotech/GM Crops: 2012', ISAAA website <http://www.isaaa.org/resources/publications/briefs/44/highlights/default.asp>.

On feeding nine billion

G. Conway, *One Billion Hungry: Can We Feed the World?* Ithaca, NY: Cornell University Press, 2012.

G. Federico, *Feeding the World: An Economic History of Agriculture, 1800–2000*, Princeton, NJ: Princeton University Press, 2005.

C. Juma, *The New Harvest: Agricultural Innovation in Africa*, Oxford: Oxford University Press, 2011.

P. McMahon, *Feeding Frenzy: The New Politics of Food*, London: Profile Books, 2014.

"牛津通识读本"已出书目

古典哲学的趣味	福柯	地球
人生的意义	缤纷的语言学	记忆
文学理论入门	达达和超现实主义	法律
大众经济学	佛学概论	中国文学
历史之源	维特根斯坦与哲学	托克维尔
设计，无处不在	科学哲学	休谟
生活中的心理学	印度哲学祛魅	分子
政治的历史与边界	克尔凯郭尔	法国大革命
哲学的思与惑	科学革命	民族主义
资本主义	广告	科幻作品
美国总统制	数学	罗素
海德格尔	叔本华	美国政党与选举
我们时代的伦理学	笛卡尔	美国最高法院
卡夫卡是谁	基督教神学	纪录片
考古学的过去与未来	犹太人与犹太教	大萧条与罗斯福新政
天文学简史	现代日本	领导力
社会学的意识	罗兰·巴特	无神论
康德	马基雅维里	罗马共和国
尼采	全球经济史	美国国会
亚里士多德的世界	进化	民主
西方艺术新论	性存在	英格兰文学
全球化面面观	量子理论	现代主义
简明逻辑学	牛顿新传	网络
法哲学：价值与事实	国际移民	自闭症
政治哲学与幸福根基	哈贝马斯	德里达
选择理论	医学伦理	浪漫主义
后殖民主义与世界格局	黑格尔	批判理论

德国文学	儿童心理学	电影
戏剧	时装	俄罗斯文学
腐败	现代拉丁美洲文学	古典文学
医事法	卢梭	大数据
癌症	隐私	洛克
植物	电影音乐	幸福
法语文学	抑郁症	免疫系统
微观经济学	传染病	银行学
湖泊	希腊化时代	景观设计学
拜占庭	知识	神圣罗马帝国
司法心理学	环境伦理学	大流行病
发展	美国革命	亚历山大大帝
农业		